AGING BONES

Aging Bones

❖ ❖ ❖

A Short History of Osteoporosis

Gerald N. Grob

Johns Hopkins University Press
Baltimore

© 2014 Johns Hopkins University Press
All rights reserved. Published 2014
Printed in the United States of America on acid-free paper
2 4 6 8 9 7 5 3 1

Johns Hopkins University Press
2715 North Charles Street
Baltimore, Maryland 21218-4363
www.press.jhu.edu

Library of Congress Cataloging-in-Publication Data

Grob, Gerald N., 1931–
Aging bones : a short history of osteoporosis / Gerald N. Grob.
p. ; cm. — (Johns Hopkins biographies of disease)
Includes bibliographical references and index.
ISBN 978-1-4214-1318-1 (pbk. : alk. paper) — ISBN 1-4214-1318-3 (pbk. : alk. paper) —
ISBN 978-1-4214-1319-8 (electronic) — ISBN 1-4214-1319-1 (electronic)
I. Title. II. Series: Johns Hopkins biographies of disease.
[DNLM: 1. Osteoporosis—history—United States. WE 11 AA1]
RC931.O73
616.7'16—dc23   2013028969

A catalog record for this book is available from the British Library.

*Special discounts are available for bulk purchases of this book. For more information,
please contact Special Sales at 410-516-6936 or specialsales@press.jhu.edu.*

Johns Hopkins University Press uses environmentally friendly book materials,
including recycled text paper that is composed of at least 30 percent post-consumer waste,
whenever possible.

CONTENTS

Disease is a fundamental aspect of the human condition. Ancient bones tell us that pathological processes are older than humankind's written records, and sickness and death still confound us. We have not banished pain, disability, or the fear of death, even if, on the average, we die at older ages, of chronic and not acute ills, in hospital or hospice beds and not in our own homes. Disease is something men and women feel. It is experienced in our bodies—but also in our minds and emotions. It can bring pain and incapacity and hinder us at work and in meeting family responsibilities. Disease demands explanations; we think about it and ask questions when affected by it. Why have I become ill? And why now? How is my body different in sickness from its quiet and unobtrusive functioning in health? Why, when an epidemic rages, has a community been scourged?

Answers to such ageless questions necessarily mirror and express time- and place-specific ideas, social assumptions, and technological options. In this sense, disease has always been a social and linguistic entity, a cultural as well as a biological one. In the Hippocratic era, more than two thousand years ago, physicians—and we have always had them with us—were limited to the evidence of their senses in diagnosing a fever, a gradual wasting, an abnormal discharge, or seizures. Their notions of the material basis for such felt and visible symptoms necessarily reflected and incorporated then-prevalent philosophical and physiological notions, a speculative world of disordered humors, "breath," and pathogenic local environments. Today we can call on a rather different variety of scientific understandings and an armory of diagnostic practices—tools that allow us to diagnose ailments not felt by patients and imperceptible to a doctor's unaided senses. In the past century, disease has also increasingly become a bureaucratic phenomenon, since sickness has been defined (and in that sense constituted) by formal disease classifications, screening practices, treatment protocols, and laboratory thresholds.

Sickness is also linked to climatic and geographic factors. How and where we live and how we distribute our resources all contribute to the incidence of disease. For example, ailments such as typhus, plague, malaria, dengue, and yellow fever reflect specific environments that we have shared with our insect contemporaries. But humankind's physical circumstances are determined in part by culture and climate—and especially by agricultural practices in the millennia before the growth of cities and industry. Environment, demography, economic circumstances, and applied medical knowledge all interact to create distinctly mapped distributions of disease at particular places and specific moments in time. The twenty-first-century ecology of sickness in the developed world is marked, for instance, by the dominance of chronic and degenerative illness—kidney and cardiovascular-system ailments, and cancer. What we eat and the work we do or do not do—our physical as well as our cultural environment—all help determine our health and longevity.

Disease is historically as well as ecologically specific. Or perhaps I should say that every disease has a unique past. Once discerned and named, every disease claims its own history. At the primary level, biology creates that identity. Symptoms and epidemiology, generation-specific cultural values, and scientific understanding all shape our responses to illness. Some writers may have romanticized tuberculosis—think of Greta Garbo as Camille— but, as the distinguished medical historian Owsei Temkin noted dryly, no one has ever thought to romanticize dysentery. Tuberculosis was pervasive in nineteenth-century Europe and North America and killed far more women and men than cholera did, but the former contagion never mobilized the same widespread and policy-shifting anxiety as the latter. Tuberculosis was a familiar aspect of life—to be endured if not precisely accepted. Unlike tuberculosis, cholera killed quickly and dramatically and was never accepted as a condition of life in Europe and North America. Its episodic visits were anticipated with fear. Sporadic cases of influenza are normally invisible, remaining indistinguishable among a variety of respiratory infections; waves of epidemic flu are all too visible. Syphilis and other sexually transmitted diseases, to cite another example, have had a peculiar and morally inflected attitudinal history. Some maladies, such as smallpox or malaria, have a long history; others, like AIDS, a rather short one. Some, like diabetes and cardiovascular disease, have flourished in modern circumstances; others reflect the realities of an earlier and economically less-developed world.

These arguments constitute the logic motivating and underlying Johns Hopkins Biographies of Disease. "Biography" implies an identity, a chronology, and a narrative—a movement in and through time. Once each disease entity is inscribed by name in our collective understanding of medicine and the body, it becomes a part of that collective understanding, and thus inevitably shapes the way in which individual men and women think about the symptoms they experience and their future health prospects. Each historically visible entity—each disease—has a distinct history, even if that history is not always coterminous with entities familiar to twenty-first-century physicians. The very notion of specific disease entities—fixed and based on a defining mechanism—is a historical artifact in itself. Dropsy and Bright's disease are no longer terms in everyday clinical practice, but they are an unavoidable part of the history of chronic kidney disease. Nor do we speak today of essential, continued, bilious, and remittent fevers as meaningful categories. Fever is now a symptom, the body's physiological response to a triggering circumstance. It is no longer a disease, as it had been through millennia of human history. "Flux," or diarrhea, is similarly no longer an entity, but a symptom associated with a variety of specific and non-specific causes. We have come to assume there will be a diagnosis when we feel pain or suffer incapacity—we expect the world of medicine to at once categorize, explain, and predict ailments.

But today's diagnostic categories are not always sharp edged and unambiguous, even if they exist as entities in accepted disease taxonomies. Emotional states present one kind of problem. Anxiety, sadness, and inappropriate phobias and compulsions were and always have been with us, just like fevers, loose bowels, or seizures. But when do extreme feelings and atypical behaviors become sicknesses? Fear and sadness, for example, are unavoidable—and normal—aspects of the human condition, yet both can become the core of powerful emotions that bring pain, stigma, and social incapacity. When do such felt emotions, and the behaviors they structure, transcend the randomness of human idiosyncrasy and the particularity of circumstance and become what clinicians have, over time, become willing to recognize as disease? A similar argument can be made about the excitability and attention span of the young. When does an idiosyncratic behavioral difference become pathology? And when does that excitability demand medical—and pharmaceutical—attention, under the legitimating rubric of attention-deficit hyperactivity disorder?

Aging, like the expression of emotions, constitutes another unavoidable aspect of the human condition. It is diverse and individual, yet marked and constituted by a variety of physical and mental changes. But nowadays no one dies of "old age," though we grow weaker and our faculties less keen, our organs functioning in less than optimal ways. Today we die of specifically coded ills, not of a very real and global condition called "old age." Then what—if anything—is pathological about aging? Or, asked another way, what is normal aging? Is gradual muscle weakness or impaired circulation a disease to be treated? Or are the conditions of advancing life to be accepted and managed? Such questions are particularly important in a modern world in which acute infectious disease is infrequent, and the chronic ills so characteristic of aging increasingly become a burden on society and on individuals.

Gerald Grob has addressed one of these ailments, osteoporosis. "Aging bones" perhaps describes his subject matter, but it is not everyone's bones that receive the label of "osteoporosis," only those that have certain characteristics and thus are associated with the likelihood of fracture—most egregiously, spontaneous fractures of the vertebrae. In contemporary usage, osteoporosis has also become a condition linked with the risk of falls and fractures. For Internet users, the National Library of Medicine website defines osteoporosis as "a disease in which bones become fragile and more likely to fracture. Usually the bone loses density, which measures the amount of calcium and minerals in the bone." The chief risk factors, the website explains, are age and being female; roughly half of all women over 50 will experience fractures of the hip, wrist, or vertebra during their lifetime.

Clinicians, laboratory workers, and epidemiologists study and debate this condition known as osteoporosis, well aware of the contingent nature of their elusive subject. Consensus conferences, task force debates, and panels at scholarly meetings all seek to reach agreement on the nature of this ubiquitous entity and the appropriate criteria for its diagnosis. As Grob demonstrates in illuminating detail, experts understand osteoporosis to be a continuously negotiated and renegotiated entity, reflecting differences in the orientation of medical specialties, in the findings of basic sciences, and—perhaps—in sources of research support. Osteoporosis occupies a place in that uncertain but increasingly well-populated ground between disease and risk factor so characteristic of our era of medical screenings and chronic diseases.

But in society at large, osteoporosis has become a *thing*, a social entity that exists outside the bodies of any particular woman or man. It exists in our minds, in our everyday routines, and in the practice of clinical medicine. Women and men are aware of it, diagnostic tools evaluate it, and drug companies market products to manage it. Osteoporosis (and its milder version, or perhaps precursor, osteopenia) is an entity that incorporates and reflects a time-specific assortment of factors: notions about aging and menopause, basic science's understanding of the biochemistry and microarchitecture of bone absorption and formation, the availability of imaging tools to measure bone mineral density, pharmaceutical marketing strategies, and epidemiological data recording the incidence of fractures. The entity that we call osteoporosis has many parents, as well as many stakeholders and coconspirators in the enterprise of maintaining it as a clinical and bureaucratic reality—as a nemesis immanent in our aging bones.

Charles E. Rosenberg

For most of my career I have been preoccupied with writing about the history of mental health policy and psychiatry in the United States. Given my interest in the history of medicine, in the 1970s a dean at my university suggested that perhaps I could teach a course on the history of the relationship between health and the environment in America. Preparing a new course inevitably influences one's thinking about the subject. In the process of teaching this course, I found my attention shifting to the history of disease, which—although related to medicine—has a somewhat different emphasis. After teaching this course for roughly twenty-five years, my thinking about disease and medicine has broadened in fundamental ways. I therefore shifted the focus of my research, and the result was a book that detailed the changing patterns of morbidity and mortality in the United States from the nation's beginnings to the present and discussed the reasons for such changes.

This work, in turn, served to intensify my ongoing fascination with the manner in which diagnoses had been created and why so many disappeared from the medical lexicon. My earlier work on the history of psychiatry had already stimulated my interest in nosology (the classification of disease). Intrigued with the manner in which psychiatric diagnoses came into existence and how they changed over time, I soon came to the realization that diagnostic categories are neither self-evident nor given. On the contrary, they emerge from the crucible of human experience; change and variability, not immutability, are characteristic. Debates about psychiatric categories continue to be phrased in scientific and medical language. Yet the fact of the matter is that these delineations are shaped by a variety of factors: the social origins and ideological, political, and social commitments of psychiatrists; the characteristics of their patients; the nature and locations of psychiatric practices; and especially the broader social, scientific, and intellectual currents prevalent at a given time.

The absence of evidence linking anatomical changes and behavior led psychiatrists to identify mental disorders by observing external signs and symptoms. This was not unique to psychiatry but was and is characteristic of medicine generally. In the absence of a nosological system based on etiology (e.g., the specific germ theory of disease), for much of the nineteenth century most practitioners defined pathological states in terms of external and visible signs, such as fever. To infer pathology by focusing on signs created formidable medical and scientific problems. For example, in the nineteenth century physicians were preoccupied with differentiating among various fevers (e.g., remittent fever, intermittent fever, scarlet fever, yellow fever, rheumatic fever). Fever, after all, is a general response to disease and not a diagnostic category. Yet there was little disposition to question diagnoses that distinguished among fevers, if only because there was no viable alternative. The problem of defining diagnostic categories persists down to the present. Consider, for example, a host of diagnostic categories for which no biopathology is evident (e.g., fibromyalgia, restless leg syndrome, chronic fatigue syndrome, irritable bowel syndrome), although they are given names as if they constitute distinct and unchanging entities. I also became fascinated with the modern preoccupation with risk factors and the tendency to define pathology in numerical terms (e.g., hypertension, glucose levels, cholesterol levels).

While doing research for articles dealing with fibromyalgia, peptic ulcers, and risk factors, I happened to peruse several medical textbooks published in the 1940s. I was already familiar with Fuller Albright's work on postmenopausal osteoporosis, which had appeared at the beginning of that decade. I was somewhat surprised, however, by the absence of the diagnosis of osteoporosis in the textbooks of that decade. The number of articles on osteoporosis published during the 1940s and 1950s was extraordinarily small. Several decades later, it was common to refer to the worldwide "epidemic" of osteoporosis. By the beginning of the twenty-first century, virtually every older woman and man was urged to undergo bone mineral density tests, and pharmacotherapy was advised for an overwhelming majority of them.

If osteoporosis posed a major threat to health (largely that of women), why did it take so long for physicians to recognize it? Was it because of their myopia or stupidity? Was it because of the absence of a technology that could illuminate its pathology? Was it because there had been little research on bone pathology? Was it because of a failure to recognize the epidemiology of vertebral and hip fractures? Intrigued by these and other questions, I decided to undertake a historical study that might shed light on them. My objective was to analyze how the normal aging of bones was transformed into a medical diagnosis that eventually included virtually all aged persons.

The term "osteoporosis" was derived from the Greek *osteon* (bone), to which was added *poros* (little hole). Thus osteoporosis meant "porous bone." There was general recognition that porous bone was a function of the aging process, but it was not necessarily pathological. Those fortunate enough to live to an old age understood that growing older inevitably led to a decline in physical and mental health, but during the latter half of the nineteenth century, the term "senescence," which in its origins simply referred to old age, had acquired a pejorative meaning that invested old age with negative connotations. With regard to osteoporosis, nineteenth-century medicine had begun to identify the manner in which bone remodeling took place, but rarely, if ever, was there an implication that porous bone represented a pathological state.

The origin of the modern diagnostic category of osteoporosis dates back to the 1920s, when endocrinologists and physiologists began to illuminate bone development and to understand the role of estrogen in strengthening bone. This, in turn, reinforced the belief that menopause was not a natural stage in women's lives, but rather represented a deficiency disease that could be treated by estrogen therapy. From this new definition of menopause, just a short leap was needed to define osteoporosis as a disease caused by estrogen deficiency. By 1940 Fuller Albright had identified postmenopausal osteoporosis as an advanced, painful, and debilitating condition, but one that affected only very small numbers of women. Albright's patients, all of whom were under the age of 65, were largely incapacitated and were forced to use corsets, braces, crutches, and canes. Albright also distinguished between postmenopausal osteoporosis and senile osteoporosis (the loss of bone that was an inevitable part of the aging process).

Interest in bone physiology intensified between the 1940s and 1960s. Researchers debated whether the loss of bone mass was a natural concomitant of the aging process or a pathological condition. The inability to distinguish between normal and abnormal bone mass density ensured that osteoporosis would remain a contested category.

During and after the 1970s, however, osteoporosis underwent a dramatic transformation. A culture of youth, a belief that the effects of aging could be postponed, and a faith in the redemptive abilities of medicine created an environment in which osteoporosis could emerge as a major threat to the health and well-being of women. Osteoporosis was identified overwhelmingly as a condition that affected women, which also reflected the prevailing Victorian model of femininity, where the ovaries were governed by the nervous system. This, in turn, seemed to confirm the beliefs that in women, emotions prevailed over rational faculties, and that females were weaker and more vulnerable than their male counterparts. Such views were widely held in a profession dominated by men.

A series of developments within the medical establishment reinforced public concern and fears. A new imaging technology made it possible to measure bone mineral density. Although the meaning of these measurements, which were expressed in numerical scales, remained obscure, this technology stimulated an interest in screening. At the same time an international community of osteoporosis researchers and clinicians redefined the diagnosis of this condition as a disease of major proportions, rather than a physiological process relating to age. National and international private and public organizations and foundations worked assiduously to publicize its dangers. The pharmaceutical industry quickly emerged as a dominant force. It marketed a variety of drugs that might prevent or treat osteoporosis; it subsidized clinical trials, researchers, and clinicians; it labored to raise public awareness about a presumably underdiagnosed and undertreated condition; and it provided financial support for epidemiological studies that framed exaggerated prevalence estimates. By the beginning of the twenty-first century, the diagnostic boundaries of osteoporosis had expanded to include virtually the entire aged population, including large numbers of individuals in their fifties and sixties. The feminist movement, which became increasingly important during the latter part of the twentieth century, posed challenges that began to transform previous interpretations about the biological character of women (including menopause), as well as proffered hostile critiques of osteoporosis. Nevertheless, osteoporosis continued to occupy a prominent role in the pantheon of diagnostic categories.

On several occasions I presented my findings on the history of osteoporosis before medical audiences. With an occasional exception, most of those in attendance proved to be less than receptive. It was difficult for them to accept an analysis that raised questions about the validity of a diagnostic category that seemed to be a given, the usefulness of screening, or the effectiveness of pharmacotherapy. Nor did they accept my claim that the influence of the pharmaceutical industry on medical practice often resulted in harm rather than good. Many expressed a belief that a partnership between pharmaceutical firms, researchers, and clinicians could only benefit both patients and their physicians. While most medical audience members professed to enjoy history, they clearly were discomforted by a presentation that offered a view of the past that had particular relevance—if largely in negative form— for the present.

To my way of thinking, history is a discipline that raises inconvenient perspectives. Too often we forget that the history of medicine is not a history of inevitable progress. Rather, it is a history in which failures far exceed successes. Witness the fact that most therapies—including recent ones—have disappeared from the medical armamentarium because they either proved to be ineffective or dangerous. To reify medicine and to promise future advances that will "conquer" disease is a dangerous illusion, and history can serve as an effective antidote. In so arguing, I recognize the very real contributions of medical science to humanity. But I am also aware of the risks that the practice of medicine entails. I hope, therefore, that readers will not dismiss this history of osteoporosis, but rather look carefully at the data and the interpretation I have presented and thus judge for themselves whether my analysis is justified.

ACKNOWLEDGMENTS

I have benefitted from comments from and discussions with a number of colleagues. Margaret Humphreys, then editor of the *Journal of the History of Medicine and Allied Sciences,* provided a shrewd analysis when I first submitted an article on this subject. Kathleen Capels did a superb job of copyediting the manuscript and making it much better than it might have been. Jacqueline Wehmueller, my editor at Johns Hopkins University Press, offered encouragement and insightful comments. My colleagues at the Institute for Health, Health Care Policy and Aging Research at Rutgers University provided an ideal environment in which history was an integral component in an organization devoted to cross-disciplinary studies. I am especially indebted to David Mechanic, director of the Institute and very close friend. For nearly four decades I have benefitted from his scholarship and wisdom. In many discussions with Allan V. Horwitz, my colleague at the Institute and collaborator on a number of projects, I have learned much about the ways in which diagnostic categories have been created and the uses to which they have been put.

AARP	American Association of Retired Persons
AHRQ	Agency for Healthcare Research and Quality
BMD	bone mineral density
BMI	body mass index
BMU	bone multicellular unit
CDC	Centers for Disease Control and Prevention
CHD	coronary heart disease
DPA	double photon absorptiometry
DXA	dual-energy x-ray absorptiometry
EFFO	European Foundation for Osteoporosis
EPOS	European Prospective Osteoporosis Study
ERT	estrogen replacement therapy
EVOS	European Vertebral Osteoporosis Study
FDA	Food and Drug Administration
FIT	Fracture Intervention Trial
FOSIT	Fosamax International Trial Study Group
FRAX	fracture risk assessment tool
GLOW	Global Longitudinal Study of Osteoporosis in Women
HDL	high-density lipoprotein
HERS	Heart and Estrogen/Progestin Replacement Study
HRT	hormone replacement therapy
IOF	International Osteoporosis Foundation
LMP	last menstrual period
MEDOS	Mediterranean Osteoporosis Study
NHANES	National Health and Nutrition Examination Survey
NHDS	National Hospital Discharge Survey
NIA	National Institute on Aging

NICE   National Institute for Health and Clinical
            Excellence
NIH   National Institutes of Health
NOF   National Osteoporosis Foundation
NORA   National Osteoporosis Risk Assessment
RCT   randomized controlled trial
SD   standard deviation
SOF   Study of Osteoporotic Fractures
SPA   single photon absorptiometry
USPSTF   U.S. Preventive Services Task Force
WHI   Women's Health Initiative
WHO   World Health Organization

CHAPTER ONE

History and Demography

In mid-twentieth-century America, few physicians would have predicted that within decades the modern diagnostic category of osteoporosis would emerge and include millions of Americans, predominantly older women. Prior to World War II, popular attitudes were shaped by a belief that the declining physical and mental health of older persons was irreversible, and that this population had little to contribute. Nor did the medical armamentarium offer much that was of use for coping with the chronic and long-duration illnesses of this group. As late as the 1960s, with a few exceptions, no dramatic discoveries had been made in the field of geriatrics.[1] With regard to bone health, little was known about the physiological processes that shaped bone resorption (where osteoclasts break down bone), and the subject was of little interest to clinicians or researchers. Yet the growing number of Americans surviving to the age of 65 and beyond was permeating the consciousness of the nation. The increase in their numbers, their emergence as a self-conscious group with distinct interests, and their rejection of the pejorative concept of senescence made them a force that could not be ignored. How would this group manage the vicissitudes of aging? What resources would enable them to survive? How would their pressing health needs be met? The answers to such questions still lay in the near future, although the passage of the Social Security Act of 1935 was a portent of change. To understand the emergence of osteoporosis, therefore, requires an understanding of both developments in medical science and those social, intellectual, economic, demographic, and political changes that would transform American society in the decades after World War II.

BONES IN HISTORY

How did bone loss—a natural concomitant of aging—become pathologized? Interest in both the aging process and the nature and character of bones dates from antiquity, yet neither were defined in pathological terms. Although a long life was highly valued, aging was simply an inescapable feature of human existence. Yet longevity had its dark side. The vigor of youth was often followed by illnesses that created disabilities and diminished mental capabilities. Nowhere was the tragic nature of aging better expressed than in Shakespeare's "Sonnet 73." In this poem Shakespeare alluded to the ravages of time on one's body and the anguish of impending death.

> That time of year thou mayst in me behold
> When yellow leaves, or none, or few, do hang
> Upon those boughs which shake against the cold,
> Bare ruin'd choirs, where late the sweet birds sang.
> In me thou seest the twilight of such day
> As after sunset fadeth in the west,
> Which by and by black night doth take away,
> Death's second self, that seals up all in rest.
> In me thou seest the glowing of such fire,
> That on the ashes of his youth doth lie.
> As the death-bed whereon it must expire,
> Consum'd with that which it was nourish'd by.
> This thou perceiv'st, which makes thy love more strong,
> To love that well which thou must leave ere long.[2]

Even the ancient world recognized that aging was accompanied by bodily changes. Nevertheless, physicians manifested little concern with the health of older persons, who (until recent times) constituted a relatively small proportion of the population. Moreover, there was little in the medical armamentarium that could arrest the inevitable process of aging and its accompanying ills and disabilities. Nor were there many diagnoses to describe the ills of older persons. Old age itself was generally offered as the cause of death.

The absence of specific diagnostic categories did not mean that bones were of little interest in the ancient or early modern world. Bone, as a matter of fact, aroused considerable curiosity. How was it created and what accounted for its differentiation from flesh? What was the relationship between bones and health? A variety of answers were given to such questions. There was a widespread belief, for example, that the oil of the marrow moistened the bone, which in turn gave rise to good health. The Book of Proverbs noted that "a merry heart is good medicine, but a broken spirit drieth the bones." Similarly, the Book of Job observed that "his pupils are full of milk, and the marrow of his bones is moistened."[3] The belief that bone health was maintained by the marrow persisted well into the eighteenth century.

Since ancient times, interest in the origins and nature of bone formation was also common. For the Greeks, heat played a central role. Aristotle emphasized the role of "seminal residue" and heat in forming bone. Plato followed by noting that bone was constructed by a god. Having sifted earth "till it was pure and smooth, he kneaded it and soaked it with marrow; then he plunged the stuff into fire, next dipped it in water, and again in fire and once more in water; and by thus transferring it several times from one to the other he made it insoluble by either."[4] Galen later echoed those beliefs.

These views persisted for many more centuries. But technology —which often plays a role in shaping human understanding— provided the impetus for change. The invention of the microscope in the 1600s and its improvement by Dutch scientist Antonie van Leeuwenhoek was a case in point. Leeuwenhoek (who also first identified single-cell organisms) described both bone as a tissue and the canal system in bone. The possibility of a new understanding of bones was hastened by the work of William Harvey, the famous English physician who identified the systemic circulation of the blood in the body by the heart, and Clopton Havers, an English physician who focused on the microstructure of bones and described the Haversian canals. Yet older conceptions, dating back to ancient Greece, still held sway. Havers, for example, believed that bone itself grew in length and diameter by interstitial (a space that intervenes between structures) expansion.

The eighteenth-century emphasis on observation and experiment, together with the dissemination of printed matter, created an intellectual and scientific environment that fostered new ways of thinking about bone. In a work published posthumously in 1743, nearly a half century after his death, Marcello Malpighi described bone as having a fibrous matrix and growing at its outer surface like a tree. One of the most fundamental contributions came from John Hunter, a Scottish surgeon and scientist. In his experiments and observations, he came to the conclusion that bones grew by outer deposition and inner resorption. To put it another way, bones—hitherto regarded as static—underwent constant change. Although Hunter could only speculate about the nature of bone resorption, his contribution helped shape future work.[5]

Building on the work of Hunter and some of his predecessors, nineteenth-century investigators created a framework for understanding the growth and development of the human skeleton. They identified both the osteoblast (the cell responsible for bone formation) and the osteoclast (the cell responsible for bone resorption), as well as the body's process of bone remodeling (a method involving alternate apposition and removal on both the inner and outer surfaces in order to shape the bone). In the twentieth century the mechanisms of bone formation and elaboration would be further elucidated by the rise of specialties such as endocrinology and orthopedics.

Oddly enough, the literature that dealt with the nature and formation of bones rarely, if ever, dealt with old age. The term "osteoporosis" did not appear until the early nineteenth century. It was coined by French pathologist Jean G. C. F. M. Lobstein in the 1820s in the context of osteitis (inflammation of the bone). It was derived from the Greek *osteon* (bone), to which he added *poros* (little hole). Thus osteoporosis meant "porous bone," a condition that Lobstein had observed in patients.[6]

Nor was age related to pathological bone structure. Some medical practitioners noted that younger and older bones were dissimilar. Astley Cooper (who pioneered several surgical procedures) observed that the bones of older people became "thin in their shell and spongy in their texture."[7] From time to time physicians offered case histories of such persons with diseased bones. In one instance a 60-year-old male who had died had severely enlarged bones; he was described as having "osteoporosis, or spongy hypertrophy of the bone." When the patient, while he was still living, was stripped and assisted to stand in an erect position, "his figure resembled much more that of an ape than of a man; he had become shorter in stature; his legs were somewhat bent; his arms fell in front of him; his head dropped forward; his chest was narrow, pelvis contracted, and belly protuberant; and thus all the characteristics of a manly figure were lost."[8] In another case a physician determined that a woman more than 90 years of age had sustained a large number of non-traumatic fractures that the patient had not noticed. The etiology (causes and origins) of the bone disorder in both cases, however, remained unknown.[9]

The term osteoporosis appeared occasionally in nineteenth-century British and European (but rarely in American) medical literature. Nevertheless, it did not denote a diagnostic category related to the aging process. In 1877 Julius Wolff (whose "law of bone adaptation" stipulated that the nature of the structure and shape of bone continually adapted to conditions that placed loads on it) described changes in adult extremities and disturbances of growth and trophic changes in children that followed infectious arthritis or joint resection. Eight years later Gustav Pommer, a German pathologist, distinguished rickets (a softening and weakening of the bones) and osteomalacia (a calcium deficit in the bone tissue itself) from osteoporosis (fully calcified and normal bone tissue, but too little of it). Subsequently, Paul Sudeck described acute reflex atrophy of bone and established it as a clinical entity. In the early twentieth century the diagnosis of post-traumatic osteoporosis emerged. In reviewing the literature that led to this diagnostic category, two French physicians concluded that "true osteoporosis is the direct result of hyperaemia produced by vasomotor changes that result from reflexes which originate in the traumatized area." In other words, the trauma altered the diameter of blood vessels and led to an increase in blood flow. Its symptoms included a loss of motor function, changes observable on x-rays, the coexistence of vasomotor disturbances, and great pain. Unlike modern definitions, in which bone loss precedes traumatic fractures, post-traumatic osteoporosis resulted from a traumatic injury.[10]

PERCEPTIONS OF AGING

Changes in disease definitions do not occur in a vacuum, and the modern diagnostic category of osteoporosis is no exception. Its emergence was shaped by changing attitudes and perceptions of the aging process; medical and scientific thinking about the relationship between aging and health; interpretations of gender; a culture that increasingly venerated youth and devalued older persons; a demographic transition that increased the percentage of older persons in the population; and a faith that new therapies, notably surgery and drugs, could arrest the undesirable attributes of aging and maintain physical function.

Historically, old age was believed to begin somewhere between 60 and 70, although some suffered the ravages of aging even earlier. In seventeenth- and eighteenth-century America the proportion of people over the age of 60 was, at least by contemporary standards, low. Unfortunately, national data are lacking for these years. Local data, however, suggests that older persons (60 and over) constituted perhaps 4–5 percent of the total population. The highest proportion was found in New England and the Middle Atlantic colonies, and the lowest in the South. As late as 1726, the renowned Massachusetts minister Cotton Mather estimated that "scarce three in a hundred live to three-score and ten." In 1830 the proportion of whites in the population over the age of 60 was only 4 percent.[11]

Old age was highly respected in colonial America, perhaps because it was atypical. Puritan cosmology also played a significant role. Old age was not an accident; rather, it was a special gift of God. "If any man is favored with long life," noted Increase Mather, "it is God who has lengthened his days." Respect for age, moreover, was not confined to New England; it was found in other colonies as well. Yet the prevailing veneration of older persons was not solely a function of religion, nor was it freely given. In an agricultural society, landed wealth was the key to power and authority. Concerned for their security in old age, parents tended to retain control over their land until the end of their lives. Children, therefore, depended on their parents until well into adulthood. Land, in other words, was deployed in a manner that reinforced the authority of older persons; this authority was also manifested in the large number of elderly men elected to positions of political power. Age by itself, however, did not guarantee respect. Those who lacked family ties or property and were unable to work because of poor health were often labeled as superannuated.[12]

To be venerated and respected might have been a source of satisfaction to those living into their sixties and beyond. Nevertheless, old age often brought with it declining health, at a time when there were relatively few means of assuaging its symptoms. Perhaps nowhere were the ravages of advanced age better illustrated than in the lives of John Adams and Thomas Jefferson. Colleagues during the struggle for independence, their relations became strained during Adams's presidency. They later reconciled and began a remarkable correspondence that can best be described as magisterial.

Until about 1818, both of these men enjoyed relatively good health. Ruminating about the life cycle in 1816, Jefferson expressed doubt about life in its later stage. "For, at the latter period [over age 60], with most of us, the powers of life are sensibly on the wane, sight becomes dim, hearing dull, memory constantly enlarging it's frightful blank and parting with all we have ever seen or known, spirits evaporate, bodily debility creeps on palsying every limb, and as faculty after faculty quits us, and where then is life?" Two years later Jefferson was "severely indisposed" by periodic attacks of rheumatism. Jefferson's sojourn at Warm Springs produced "imposthume, general eruption, fever, colliquative sweats, and extreme debility." By 1820 his legs were swollen and he could "walk but little." In subsequent years writing was physically painful, although Jefferson's correspondence with Adams made him forget the "hoary winter of age, when we can think of nothing but how to keep ourselves warm, and how to get rid of our heavy hours until the friendly hand of death shall rid us of all at once." Adams remained in better health than Jefferson, but he, too, noted his declining vision and trembling hands. At the age of 87 Adams could no longer mount his horse, but added, perhaps whimsically, that he was able to walk three miles over "a rugged rockey Mountain." Death was not an evil; it was rather "a blessing." Despite their declining health, the two men rarely obsessed over their growing frailties, and their minds remained as acute as ever. Six months before he died, Adams wrote that he contemplated death "without terror or dismay." Adams's and Jefferson's correspondence ended three months before their deaths, which occurred on July 4, 1826—the fiftieth anniversary of the signing of the Declaration of Independence. Jefferson was 83 and Adams 91.[13]

In the late eighteenth century, perceptions of aging began a remarkable transformation that ultimately reshaped popular views. Rather than being revered as a source of wisdom, older persons began to be increasingly depicted in negative terms and their value to society questioned. The causes of this change were complex, but when the process had run its course, older patterns of deference had largely disappeared.

In the mid-eighteenth century the ideas of equality and freedom—subsequently so ably described by Tocqueville in his classic *Democracy in America*—began to undermine older, hierarchical conceptions of the social order. The idea that all men (except slaves and women) had an inalienable right to liberty undermined the older belief that individuals were part of a cohesive community in which obligations and ties between generations were indissoluble. Certainly the challenge to imperial authority during the revolutionary decades fostered a faith in the virtues of individual liberty, which slowly disrupted the social coherence that was so characteristic of colonial America.

At the same time, the growth of a market economy not only promoted social fragmentation, but also created an increasingly unequal distribution of wealth. While land was still valued, opportunities in the commercial sphere offered young men a chance to strike out on their own and not be dependent on their parents. The resulting emergence of a market-based elite and an increase in economic inequality weakened community ties, heightened social tensions, reinforced class distinctions, and undermined the idea of reciprocal obligations between generations. Moreover, wealth, particularly in the Middle Atlantic and northern states, tended to be concentrated in urban areas, where the bonds of kinship and community grew even more tenuous. People began moving from one region to another, further weakening community cohesion.

Other developments contributed to the loosening of bonds between parents and children. In colonial America large families were the norm, rather than the exception. As late as 1800, the number of children per completed marriage was about seven, and women, on average, bore children for about seventeen years. Under these circumstances most parents had children and then young adults in their homes until the parents were into their sixties. But as the number of children per household fell in the nineteenth century, most couples found themselves alone in their later years. The dividing line between generations became increasingly distinct, especially in growing urban areas. Poor families, in particular, felt the stress of social isolation, as parents lacked the resources to maintain themselves in their older years. Industrial changes during the late nineteenth century further undermined the status of older Americans. Labor-force participation by older males began to decline, and widowhood only fostered greater dependence.[14]

In colonial and revolutionary America, there was a pervasive belief that the presence of poor and dependent persons was part of a divine plan, and that society had a moral obligation to alleviate their condition. This was as true for the aged as it was for other victims of misfortune. In the nineteenth century, the aged were placed in a separate category, and senescence was increasingly interpreted in pejorative terms. Dependent children and unemployed young adults could be reformed and become productive citizens. The aged, however, represented a hopeless class and thus could contribute little of value to society.

DEVALUING OLD AGE

Nowhere was the negative perception of the elderly better portrayed than in the writings of social analysts and physicians. The very terminology they and others employed in the nineteenth century was suggestive. From the fifteenth to the eighteenth centuries, terms such as "senile," "senility," and "senescence" had no pejorative meaning; they simply referred to a state of being old. By the mid-nineteenth century, these words were equated with weakness, and by the end of the century they implied a pathological state. It was not an accident that such usage appeared when it did; it paralleled the changing status of older persons.[15]

A rising faith in the paramount importance of classification and the collection of statistical data was especially important in reshaping attitudes toward aging. By this time several currents had converged to give rise to a type of social inquiry whose methodological distinctiveness was a commitment to quantitative research. The seventeenth-century mercantilist concerns with population and vital statistics were reinforced by nineteenth-century Baconian science, which tended to identify all of science with taxonomy. To this was added the fascination with social problems characteristic of most modern nations. This fascination, in turn, stimulated interest in quantitative methods, to a degree where virtually all significant problems were defined and described in statistical terms. Underlying the application of a quantitative methodology was the assumption that such a procedure could illuminate and explain social phenomena. Consequently, early and mid-nineteenth-century medicine and science were preoccupied with the development of elaborate classification systems capable of ordering a seemingly infinite variety of facts.[16]

The importance of age categories and the belief that it was possible to develop general laws from these categories was especially evident in the work of Benjamin Gompertz. Before then, the collection of mortality statistics remained an end in itself; little or no thought was given to patterns or explanations. Although Gompertz was a practicing English actuary, he went far beyond the problem of providing estimates of premiums for life annuities. His concern was with the development of a law of human mortality, which led him to publish a famous and influential article in 1825. Simply put, Gompertz—similar to Malthus—noted that age increases arithmetically, while mortality progressed geometrically. Death was not a random event, but increased by about 10 percent per year from adulthood to old age. To deal with the fact that not all individuals lived to similar ages, he distinguished between chance—"without previous disposition to death or deterioration"—and what he called "a deterioration, or an increased inability to withstand destruction." Human beings, in other words, had a more-or-less fixed span of life, although he could not spell out its precise limits.[17]

Adolphe Quetelet (a Belgian statistician) also played a key role in dividing humans into age categories and then elaborating on their capabilities. Influenced by the physical sciences, Quetelet hoped to use its methods to develop what he called "social mechanics," or "social physics." Mathematical analysis of social facts could produce generalizations. To concentrate on individuals, however, would preclude the generalizations so characteristic of the sciences. Quetelet therefore strove to examine large numbers of individuals in order to calculate averages. In 1831 he published his *Research on the Propensity for Crime at Different Ages*. Quetelet acknowledged that it was impossible to predict the behavior of any single individual; accidental causes played a crucial role. But by focusing on aggregate behavior, one could begin to deal with social (and therefore constant) causes. Out of this came Quetelet's concept of the "average man." In his book, Quetelet attempted to determine the propensity for crime at different ages. His conclusion, for his era, was striking. "Among all the causes which have an influence for developing or halting the propensity for crime, the most vigorous is, without contradiction, age." The highest crime rate for men occurred around age 25, when their "intensity of physical strength and passions" peaked. The result was violent crime, whereas in later life cunning was substituted for strength. Crimes committed by women, however, peaked at age 30 and were committed against property, rather than persons.[18]

In his major work, the *Treatise on Man and the Development of his Faculties* (written in French in 1835 and translated into English in 1842), Quetelet elaborated on his concept of the "average man." Once again, age was central to his analysis. In this work he divided large numbers of subjects according to age, in the hope of establishing "correct average proportions" for each stage of life. Previous writers, Quetelet noted, "have not determined the age at which his [man's] faculties reach their maximum or highest energy, nor the time when they commence to decline. Neither have they determined the relative value of his faculties at different epochs or periods of his life." In Quetelet's magnum opus there were hints that advanced age was synonymous with decline. In discussing dramas written for French and English audiences, he found that dramatic talent was evident between ages 25 and 30. It increased in a vigorous manner until age 50 or 55, and then gradually declined.[19]

The very creation of an age-based classification system was of critical importance, for it heightened awareness of age as a distinct entity, capable of being categorized in a variety of ways. Although Quetelet was undoubtedly one of the most prominent figures in developing an age-based classification system, he was not alone. The concept of senescence occupied an increasingly distinct niche in medical thinking. By the late eighteenth and early nineteenth centuries, figures such as Xavier Bichet (a French anatomist and physiologist) and others conflated aging with specific bodily conditions. By the middle of the century, French clinicians, in particular, began to define aging as a progressive disease that was manifested in physiological and anatomical changes. Many of them worked in hospitals that were welfare institutions, housing large numbers of elderly persons who were destitute and unhealthy. In conducting autopsies on such patients, these clinicians found numerous signs of degeneration, including arteriosclerosis, ossification, and calcification. Working with this type of patient population led to the emergence of a geriatric medicine that united concepts of senescence with decline. In Germany, Theodor Schwann and Rudolf Virchow called attention to the cell as the basic unit of life, which ultimately provided new insights into pathological processes.[20]

During the latter half of the nineteenth century, the belief in senescence grew even more pronounced. Many American physicians, together with their English brethren, identified what they called the "climacteric period." The term was derived from the Greek *climacter* (a step, staircase, or ladder) and used to designate a dangerous period of life. The male's climacteric was analogous to the female's menopause; both appeared between ages 40 and 50. Among females, menopause was often followed by varieties of mental disorders, including melancholia, hallucinations, and suicidal tendencies. "While there is no specific form of mental disorder that can be properly termed 'climacteric insanity,'" wrote George Rohé, "there can be no doubt that the menopause must be considered as one of the exciting cause[s] of mental disease."[21] Among males, the symptoms and signs of the climacteric were similar to those of the female, although they were sometimes mild and often proceeded gradually. Nevertheless, functional diminution, due to progressive degenerative changes, began to become noticeable. "The salient point to remember," noted one physician, is "that the highly civilized man . . . reaches the apex of life—his 'climacteric'—somewhere around the fifties," at which time an inevitable decline in body and mind followed.[22]

The climacteric and decline played a central role in the writings of Ignatz L. Nascher, a physician who coined the term "geriatrics" from the Greek *geras* (old age) and *iatrikos* (relating to the physician). In his view the normal cycle of life fell into three distinct periods—development, maturity, and decline—with each one lasting about thirty years. The phase designated as maturity was fractured in the middle by menopause for the female and corresponding changes for the male. The senile climacteric period for both sexes began at about age 70, when a series of degenerative changes appeared. The signs of such degeneration included weakened intellectual faculties, mental depression and apathy, a diminution of strength, and a susceptibility to heart disease and arteriosclerosis, to cite only a few. Senility, according to Nascher, was a normal physiological phenomenon at a time of life when degeneration and decay were natural and inevitable and could only be halted by death. His description of the aging process only reinforced negative stereotypes of the elderly.[23]

Thus decrepitude and old age went hand in hand. The causes, however, remained a matter of opinion. Dr. Charles-Édouard Brown-Séquard, for example, attributed senility to "diminishing action of the spermatic glands," whereas Elie Metchnikoff attributed it to poisonous microbes in the intestines that produced toxic reactions.[24] In a survey of explanations about the cause of old age, one journalist wrote that "we know practically nothing whatever." Nevertheless, there was general agreement that surviving to old age was not necessarily a virtue.[25]

The devaluation of age and the negative stereotypes of older persons that were intrinsic to American society in the nineteenth century and much of the twentieth had other roots as well. The geographical boundaries of the nation were expanding, its population was growing rapidly, technology and industrial development were transforming people and institutions, and wars resulted in the acquisition of domestic and overseas territories. All of these fostered the view that the United States was a young and vigorous world power. Change, rather than continuity, seemed to be the norm. In such a society youth—not old age—came to be venerated. A "cult of youth"—to use David H. Fischer's words—came to dominate American society during the last two centuries. Even older Americans sought to emulate younger people in terms of their dress, physical appearance, and behavior.[26]

In the writings of Transcendentalist figures, contempt for age was common. "Nature abhors the old," wrote Ralph Waldo Emerson, "and old age seems the only disease; all others run into this one." "I have lived some thirty years on this planet," noted Henry David Thoreau, "and I have yet to hear the first syllable of valuable or even earnest advice from my seniors. They have told me nothing and probably cannot teach me anything."[27]

During the second half of the nineteenth century, a novel concept of aging became popular. Known as the "fixed period," it repudiated the visions of earlier sanitary reformers who believed that many diseases were preventable by human agency. The fixed-period concept was popularized by such figures as Dr. George M. Beard (who also developed the diagnosis of neurasthenia). When pursuing medical studies at Yale in the early 1860s, Beard began to collect biographies of great figures. By computing the mean age at which they accomplished their most original contributions, he came to the conclusion that 70 percent of the work of the world was done before age 45, with the most creative being between ages 30 and 45. Inevitable decline then followed. The moral and reasoning faculties were the first to show signs of cerebral disease; their decay in advanced life led to the impairment of other faculties. Beard was especially critical of the power and authority exercised by old men. In many ways Beard's view of ageism was admirably suited to the new capitalist industrial order, for it elevated productive capacity as one of the highest social goods.[28]

Perhaps the most extreme portrayal of ageism came from Dr. William Osler, in his valedictory address at the Johns Hopkins University in 1905.

> I have two fixed ideas. . . . The first is the comparative uselessness of men above forty years of age. . . . The world's history bears out the statement. Take the sum of human achievement in action, in science, in art, in literature—subtract the work of the men above forty, and, while we should miss great treasures, even priceless treasures, we should practically be where we are to-day. . . . The effective, moving, vitalizing work of the world is done between the ages of twenty-five and forty years. . . .
>
> My second fixed idea is the uselessness of men above sixty years of age, and the incalculable benefit it would be in commercial, political, and in professional life if, as a matter of course, men stopped work at this age. . . . As it can be maintained that all the great advances have come from men under forty, so the history of the world shows that a very large proportion of the evils may be traced to the sexagenarians—nearly all the great mistakes politically and socially, all of the worst poems, most of the bad pictures, a majority of the bad novels, and not a few of the bad sermons and speeches. . . . Only those who live with the young can maintain a fresh outlook on the new problems of the world.[29]

Osler's address received national publicity. Whether he was completely serious, however, is not clear, given the fact that he had a well-developed sense of humor. In 1905, as a matter of fact, he was 56 years of age and was leaving Johns Hopkins to assume the Regius Professorship of Medicine at Oxford University, a position he held until his death in 1919.

The portrayal of senescence in negative terms and of youth in positive terms was often echoed in popular literature. In 1906 Carl Snyder published an article in *Living Age* entitled "The Quest of Prolonged Youth." Old age, he conceded, was defined as "a disease, that is to say it is essentially a pathological condition. There are not a few of the most eminent physiologists living who regard it as practically a specific disease." But was it "a wholly hopeless problem?" Snyder's answer was equivocal. "It may be that we shall never learn to avert old age. It may be, but there is no á priori certainty." Nevertheless, he believed that it was possible to learn its cause. Snyder cited the work of numerous medical figures who were working on the problem. The greatest impediment was the lack of money to promote research, a view that Snyder had advanced three years earlier when he noted the inferior position that the United States occupied in the world of science. Implicit in his 1906 article was a faith that such research could prolong youth and thus modify or prevent the disabilities that accompanied the aging process.[30]

Nonetheless, the vision of aging as decline did not go unchallenged. In 1921 G. Stanley Hall (the recently retired president of Clark University and a major figure in the history of American psychology) published an article ruminating on his departure from academic life. In it, Hall expressed enthusiasm about beginning a struggle against evils that earlier he had lacked the fortitude to attack. Among them was "the current idea of old age itself." What then was the role that aged persons could play in society? To this question Hall offered a clear and incisive answer. "But the man of the future will be ashamed and feel guilty if he cannot plan a decade or two more of activity; and he will not permit himself to fall into a thanatopsis mode of mind, or retire to his memories, or to the chimney-corner. . . . We need prophets with vision, who can inspire and also castigate, to convict the world of sin, righteousness, and judgment. Thus there is a new dispensation which gray-beards alone can usher in. Otherwise mankind will remain splendid but incomplete."[31]

The following year Hall's five-hundred-page book, *Senescence: The Last Half of Life*, appeared. It challenged the superficial view that conflated aging with regression to a form of childhood, and it rejected the prevailing devaluation of older persons and their relegation to a separate class of useless human beings. The task, Hall insisted, was to "construct a new self." While conceding that age brought changes in its wake that required a retreat from youth, it also held out the hope of an advance to a new position that compensated for what had been lost. "The function of competent old age is to sum up, keep perspective, draw lessons, particularly moral lessons."[32]

Hall's emphasis on the positive aspects of old age notwithstanding, contempt for advanced age was strengthened by the very forces that were in the process of transforming American society. This contempt was not only expressed in intellectual and medical terms; it was also mirrored in the changing status of the elderly. By the end of the nineteenth century, the movement of the American population into urban areas had separated work from the household and hastened the rise of wage labor. The emergence of large corporations and other industrial changes had a dramatic impact on the workforce. The ideal of efficiency—inherent in the bureaucratic procedures that characterized such large corporate entities—strengthened the belief that older persons constituted an undue burden. The payment of full wages to older workers who were no longer as efficient as their younger counterparts was deemed to be unjustified. Moreover, the presence of the former in the workforce had a demoralizing effect, to say nothing of the fact that they blocked the advancement of younger and presumably more able persons.[33]

The devaluation of old age, in combination with the rising emphasis on efficiency, led large-scale organizations to adopt retirement plans, in order to rid themselves of older workers. Even the term "retire" was invested with new meanings, as evidenced in the 1880 edition of Webster's *American Dictionary*. The term initially meant "to cause to retire, specifically to designate as no longer qualified for active service." To this definition of the word, "superannuate" was added: "To give pension to, on account of old age, or other infirmity." To put it another way, in the new industrial economy—with its modern, mechanized factories— older workers were burdens rather than assets, and their removal was indispensable for continued progress.[34] In urging Congress to enact legislation providing pensions for federal employees, Secretary of the Treasury Franklin MacVeagh noted that the lack of such a system inhibited "economy through efficiency." "The only thing to do is to throw the old men and old women out on the street [a practice that he regarded as inhumane], or keep them in employment with their partial efficiency."[35] MacVeagh's views were echoed by F. Spencer Baldwin (a Boston University economist). "It is well understood nowadays that the practice of retaining on the pay-roll aged workers who can no longer render a fair equivalent for their wages is wasteful and demoralizing." Until the adoption of a retirement system, municipal governments "will continue to be handicapped by the dead weight of inefficiency, resulting from the continuance in employment of large numbers of worn-out workers."[36]

DEMOGRAPHY

The devaluation of old age occurred at precisely the same time that the age structure of the population was continuing to undergo dramatic changes. In the nineteenth century Americans were relatively young. In 1850 and 1900 the median ages were 18.9 and 22.9, respectively. The proportion of elderly people was, at least by contemporary standards, relatively low. In 1850 about 4 percent of the population were age 60 and over; a half century later the figure was 6 percent. In 1995, 17 percent of Americans were 60 or older (or 43.6 million out of a total population of over 262 million).[37] The reasons for this spectacular increase are often misunderstood. Much of it was due to the rapid decline in infant and child mortality that commenced in the late nineteenth century and accelerated in the twentieth. As more and more of the younger population survived the vicissitudes of infectious diseases that previously had taken such a high toll, their chances of surviving into adulthood increased. Even in the eighteenth and nineteenth centuries, survival to age 20 meant that individuals would often live into their sixties. Medical advances admittedly played a role in the rise in life expectancy after World War II, but the crucial element, and largest decline, was in mortality in the young.

The increasing number of aged members in the population also gave rise to a new category, the very old (currently defined as those over age 85). In 1900 this group was not enumerated separately, but was included in the 65-and-older category. In 1997, the very old included more than 3.8 million persons, and the U.S. Bureau of the Census predicted that by 2050 this group could be as large as 31 million. While it is clear that the gains in longevity were not equally distributed, virtually all groups were better off than their predecessors had been at the beginning of the twentieth century.[38]

The change in the age distribution in the American population was also accompanied by new patterns of morbidity and mortality. In 1900 infectious diseases remained the major cause of mortality. Of the fifteen leading causes of death, infectious diseases accounted for 56 percent of the total, chronic and degenerative diseases for 37 percent, and accidents and suicide for 7 percent. Of all age groups, infants and young children remained the most vulnerable. The death rates for white males and white females, both under the age of 1, were 155 and 125 per 1,000, respectively; the comparable figures for non-whites (largely African Americans) were 342 and 286 per 1,000. The risk of dying was also high among those between ages 1 and 4. Such substantial mortality rates were overwhelmingly due to infectious diseases. Infants and children accounted for the majority of all deaths.[39]

For a variety of reasons, infant and child mortality in America began to decline precipitously around the turn of the century. Specific medical therapies did not play a significant role; what were more important in this decline were a range of public health activities related to sanitation, pure water, housing, quarantine, and other general preventive measures. Accurate diagnoses of infectious diseases—even if recognition of a condition did not yet lead to specific treatments for it—affected the provision of care, which in turn had a positive effect on outcomes. Consequently, more and more children survived to the age of 20, at which time they could expect to live into their sixties and beyond.[40]

As the median age of the population increased, there was a dramatic change in the causes of mortality. Since more and more people lived until their sixties and seventies, the major factors shifted to chronic diseases. In 1998, diseases of the heart accounted for 31 percent of all deaths, malignant neoplasms for 23.2 percent, cerebrovascular diseases for 6.8 percent, chronic pulmonary diseases for 4.8 percent, and diabetes mellitus for 3.9 percent. To put it another way, out of 2.34 million deaths, 1.75 million were among people age 65 and older.[41]

During the first half of the twentieth century, there was relatively little in the medical armamentarium capable of arresting the progress of most long-term, chronic, degenerative diseases. "The problem of chronic disease will not be downed," wrote George H. Bigelow and Herbert L. Lombard in a pioneering study published in 1933. "Increasingly great numbers of people are ill, crippled and dying from chronic disease. . . . There is hardly a family in Massachusetts without immediate experience with cancer, heart disease, or rheumatism. . . . Not only do chronic diseases make up two-thirds of all deaths in Massachusetts, whereas fifty years ago they were but one-third, but also from the duration as noted on the death returns there is a marked increase in the length of the chronic disease that kills." Noting that the "complete elimination of sickness and disease may not be even theoretically desirable," these authors suggested that "the span of crippling and terminal illness" be reduced to a minimum, followed by a "humane departure."[42]

The views expressed by Bigelow and Lombard were by no means atypical. In their analysis of the types of deaths among Metropolitan Life Insurance Company clients, Louis I. Dublin and Alfred J. Lotka conceded that "we have had at most moderate success" in combating the diseases characteristic of midlife and after. Cure, insisted Ernst P. Boas (of Montefiore Hospital in New York City), was an unrealistic goal. Chronic illness and disability required management and care.[43] The seeming inability to arrest most long-term illnesses only reinforced the prevailing view that old age was synonymous with decline.

Slowly but surely, diseases of old age increasingly held the attention of researchers and clinicians in the first half of the twentieth century. In the numerous medical and lay discussions before the 1940s about the relationship between aging and decline, however, there was a striking absence of publications that mentioned bone fragility. Osteoporosis simply meant a weakened, softened, or porous bone. During this period, the debate over its etiology and therapy were marked by disagreements. To some, osteoporosis was due to a traumatic injury followed by prolonged immobilization. In such cases even patients in casts were urged to walk on crutches, bearing some weight on the limb; non-use would merely worsen the condition. To others, the theory that attributed rarefaction (less density) to inactivity and a lack of functional stimulation was not supported by empirical data. According to Duval Prey and John M. Foster, many cases "which have been immobilized over a long period of time develop no areas of rarefaction, while in other cases that have never been immobilized, develop an osteoporosis." These authors agreed with René Fontaine and Louis G. Hermann in advocating that in cases of post-traumatic osteoporosis, surgical treatment—sympathectomy or periarterial sympathectomy (operations on the sympathetic nervous system)—demonstrated promising outcomes.[44] Such debates, however, were far removed from the modern diagnosis of osteoporosis.

Nowhere was the absence of the diagnosis of osteoporosis better illustrated than in two classic medical texts. The 1892 first edition of William Osler's famous and influential *The Principles and Practice of Medicine*, and later editions by him, did not even list bone disorders or osteoporosis. The fifteenth edition, in 1944, prepared by Henry A. Christian, had a section on diseases of the bone but omitted any mention of osteoporosis. The same was true of the 1924 first edition of Logan Clendening's widely used textbook.[45] Clinicians admittedly dealt with traumatic injuries among the aged, but they never attributed such injuries to an underlying pathological condition.

Interest in the physiology of bones, however, would shortly undergo a dramatic upswing. Developments in medicine, notably endocrinology, began to illuminate the process of bone resorption in older persons. At the same time, work in reproductive endocrinology began to explain the role of hormones in menopause, which was redefined not as a natural physiological process, but rather as a pathological deficiency that also hastened bone resorption in women. The isolation of estrogen, and commercial development of this hormone, held out hope that a variety of female maladies were amenable to medical intervention. Such knowledge, by itself, might have had little impact. But during the second half of the twentieth century, the research of endocrinologists was invested with new meanings by those forces that were in the process of changing institutions and modes of thinking. To understand the creation of the diagnosis of osteoporosis, we now turn to the internal history of medicine and endocrinology during and after the 1920s.

CHAPTER TWO

The Origins of a Diagnosis

The diagnostic category of osteoporosis did not appear until the mid-twentieth century, because of the weakness in and complexity of the knowledge base about bone physiology. Progress in illuminating the manner in which bones developed and changed over time would depend on the emergence of such specialties as endocrinology and technological innovation.

Nevertheless, nineteenth-century European anatomists and physiologists interested in bones had already begun to lay the foundation for an understanding of how the human skeleton grew and developed. In 1816 John Howship (a well-known English surgeon) first described the pitted and eroded characteristics of bone undergoing resorption, although cells had not yet been identified, so Howship could not describe the osteoclast.[1] The emergence of the cell doctrine during the latter part of the nineteenth century led Swiss anatomist and physiologist Albert von Kölliker to name the osteoclast in 1873 and propose its relationship to bone resorption in a classic monograph. Several years earlier Carl Gegenbauer (a German anatomist and supporter of Darwin's theory of evolution) described the osteoblast and its function. Others added to an understanding of bone composition, structure, and physiology. Still, nineteenth-century findings about bone and the skeleton, however significant, did not explain what drove the process of bone remodeling. The focus at that time was either on a causative increase in osteoclastic activity or a causative decrease in osteoblastic activity, both of which presumably operated independently of each other.

During the first quarter of the twentieth century, the causes of bone atrophy remained mysterious; speculation and eclectic but unproven theories were common. In 1929 J. Albert Key (of the Shriners Hospital and Department of Surgery at Washington University in Saint Louis) attributed bone atrophy to general causes (including senility, hunger, marasmus, biliary and pancreatic fistulas, increased metabolism, scurvy, rickets, and osteomalacia) and localized sources (such as disuse, trauma leading to bone atrophy, inflammation, neoplasms, pressure, and neurotrophic disturbances).[2] Such eclecticism, however, was already in the process of being transformed by research in specialties whose members seemingly manifested little interest in bone physiology.

THE EMERGENCE OF ENDOCRINOLOGY

A clearer understanding of the process that drove bone remodeling had to await the emergence of endocrinology, particularly the discovery of what later became known as hormones. In the nineteenth century there was considerable interest in the secretions of the sex glands and the parathyroid and thyroid glands. Concern with these secretions led to a redefinition of femininity, which heretofore had been characterized by the uterus. At midcentury an important shift began to take place. In 1848 young Rudolf Virchow gave a lecture in which he discussed the function of the ovary, which was the production of the ova. Virchow rejected humoralism (the doctrine that disease arose from the four bodily humors of black bile, yellow bile, phlegm, and blood) in favor of a solidistic perspective (the theory that disease arose in the solid components of the body). Menstruation was correlated with ovulation, and these periodic, rhythmic changes were governed by the nervous system. More importantly, the ovaries defined femininity. "The female is female," Virchow argued, because of her reproductive glands. All her characteristics of body and mind, of nutrition and nervous activity, the sweet delicacy and roundedness of limbs, . . . the development of the breasts and non-development of the vocal organ, the beauties of her hair and the soft down on her body, those depths of feeling, that unerring intuition, that gentleness, devotion, and loyalty—in short, all that we respect and admire as truly feminine, are dependent on the ovaries. Take the ovaries away and we get the repulsive, coarsely formed, large-boned, moustached, deep-voiced, flat-breasted, resentful, and egoistic virago [*Mannweib*].

Virchow, of course, was offering more than a medical statement. He was implicitly endorsing the Victorian model of femininity. This, in turn, seemed to confirm the belief that, in females, emotions prevailed over rational faculties (in contrast to the view that, in males, the intellectual propensities of the brain dominated).[3]

Virchow's formulation was by no means idiosyncratic. In the eighteenth and nineteenth centuries, gender differences were increasingly explained in terms of biological determinism. Women were smaller and weaker than men, suffered more illnesses, were more emotional, and were inferior to men in reason and intellect. In the twentieth century, psychoanalytic and psychodynamic psychiatry wedded the decline of estrogen levels to negative emotional states (notably depression). The role of women was reduced to attracting a man, producing children, and nurturing the family.[4]

Concern with sexual physiology was extended to interest in the secretions of the male testes. Claude Bernard (an eminent French physiologist and contemporary of Virchow) developed the concept of the *milieu intérieur*. Bernard suggested that as-yet-unknown substances regulated an organism's internal environment (an idea that Walter B. Cannon later named "homeostasis"). The concept of internal secretions was given impetus by Charles-Édouard Brown-Séquard who, in a paper in 1889, announced the first use of ovarian extracts. Brown-Séquard also noted that the testes contained an active and invigorating substance that would increase strength, vigor, and mental activity. Enthusiasm for what became known as "organotherapy" peaked but quickly waned.[5]

The idea of "internal secretions" drew the attention of others. In 1891 George R. Murray (an English physician who later was a pioneer in endocrinology) described the successful treatment of myxedema (severe hypothyroidism) with thyroid extract. Shortly thereafter Emil Knauer (a Viennese physician) removed the ovaries from adult rabbits and then grafted fragments from them into the same individuals at new sites. This procedure prevented the castrate atrophy that would have followed the oophorectomy. Knauer reasoned that the grafted fragments were producing something that was transported to the uterus via the bloodstream. Subsequently, Josef Halban (another Vienna-trained gynecologist) implanted pieces of ovaries from adult guinea pigs under the skin of young guinea pigs, which hastened the rapid development of the latter's uteruses. Halban proposed that ovarian secretions, traveling via the blood, were crucial for the development of female genitals. Excretions from the ovaries and other endocrine glands became known as hormones, a term coined by Ernest Starling in 1905. Starling, a British physiologist, had (with William M. Bayliss) discovered the first hormone in 1902, which they named secretin (a substance secreted by the intestinal lining to stimulate the pancreas to release digestive juices). A few years earlier epinephrine (adrenaline) had been discovered by several investigators working independently; it was synthesized in 1904. The word "hormone" came to mean a chemical released by a gland that sent out messages to other parts of the organism.[6]

By the turn of the century, endocrinology (the study of the endocrine system and hormones) was emerging as a legitimate research specialty. The importance of the endocrine glands and the hormones they produced was demonstrated by what happened when the thyroid and parathyroid glands in animals and humans were removed: the mammals suffered severe physiological consequences. In 1915 Friedrich Schlagenhaufer suggested that bone disease was the result, rather than the cause, of parathyroid hyperplasia (an enlargement of the parathyroid glands). Subsequently, the development of parathyroid extract was shown to influence serum calcium (the level of calcium in the blood).

The rapid development of endocrinology led to an enthusiasm for rejuvenation therapy. Men underwent testicular transplants in the hope of restoring their youthful vitality. Rejuvenation therapies, however, were not for females. In the late nineteenth century the popularity of removing both ovaries (double ovariotomy, or Battey's operation) for a variety of female ills, including mania and epilepsy, resulted in menopausal symptoms in young women. Ovarian extracts were frequently prescribed for such women. The enthusiasm for the alleged benefits of these therapies during the early twentieth century only enhanced the status of endocrinology as a research and medical specialty. Moreover, the use of extracts and interest in their further development made the pharmaceutical industry an important player.[7]

Research on hormones accelerated rapidly in the early twentieth century. In 1917 the formation of the Association for the Study of Internal Secretions (which in 1952 became the Endocrine Society) and the appearance of the journal *Endocrinology* symbolized the growing interest in endocrinological research. The organization was often ridiculed in its early days because of the enthusiastic claims made by proponents of rejuvenation therapies. Harvey Cushing's presidential address before the association in 1921, dealing with disorders of the pituitary gland, was a sharp indictment of endocrinology. Cushing even ignored the organization's journal and instead published his address in the *Journal of the American Medical Association* (*JAMA*), perhaps because he wished to reach a larger audience. "A good many of us, I fear," he told his colleagues,

> have completely lost our bearings in the therapeutic haze eagerly fostered by the many pharmaceutical establishments. . . . Surely nothing will discredit the subject in which we have a common interest so effectively as pseudoscientific reports which find their way from the medical press into advertising leaflets, where clev-

erly intermixed with abstracts from researches of actual value the administration of pluriglandular compounds is promiscuously advocated for a multitude of symptoms, real and fictitious. The Lewis Carroll of today would have Alice nibble from a pituitary mushroom in her left hand and a lutein one in her right and presto! She is any height desired.

Nevertheless, the organization and its journal made steady progress. The isolation of insulin by Frederick Banting and Charles H. Best in 1921 was perhaps the most spectacular finding, hastening the legitimation of endocrinology as a major medical specialty.[8]

It was not surprising, therefore, that by the end of the decade two young investigators discovered estrogen, a hormone that would subsequently play a crucial role in understanding bone development. Edgar Allen (a biologist) and Edward Doisy (a biochemist) were just beginning their careers. The former worked on the estrous cycle of mice, the latter on the purification of insulin. In chatting about the pancreas gland and the secretion of insulin (a hormone that removes excess glucose from the blood and causes liver, muscle, and fat tissue to store it as glycogen), Allen and Doisy speculated that the relationship between the ripening of follicle cells in the ovaries and developments in the uterus and vagina might be governed by hormones. In 1929 they succeeded in isolating a sample of pure crystalline estrogen, although their work was not done in splendid isolation. By the 1920s gynecologists were supplying hormone researchers with urine from pregnant women and, based on laboratory findings by pharmaceutical companies, prescribing ovarian preparations produced by these firms.[9]

Despite the Great Depression of the 1930s, pharmaceutical companies began to expand their research divisions in search of new commodities. A variety of estrogen products that were developed in their laboratories were marketed to treat menopausal women. Initially these products were available at pharmacies, without prescriptions, to all who could afford them. The passage of the Federal Food, Drug, and Cosmetic Act of 1938 sharply modified the original Pure Food and Drug Act of 1906 and its subsequent amendments. The new legislation introduced two significant changes. First, manufacturers had to apply to the government to sell new drugs. Second, to clarify enforcement of this law, the U.S. Food and Drug Administration (FDA) created a class of drugs that could only be sold by prescription. These changes were not opposed by large pharmaceutical firms, since their estrogen products could pass FDA scrutiny, while other producers would be driven out of the market.[10]

The law had significant consequences for women. It strengthened an ongoing process of dramatically reinterpreting the meaning of menopause. Menopause, although a natural biological phenomenon that all women experienced, nevertheless was linked with a host of physical and mental problems. The identification of progesterone and estrogen, however, introduced a new element. If menopause was redefined as a condition marked by estrogen deficiency, why could it not be treated with estrogen therapy? In other words, menopause, far from being part of the normal life cycle, was a deficiency state requiring medical intervention. By 1940 an alliance of researchers, physicians, and the pharmaceutical industry had begun to medicalize female aging and thus provide a rationale for its treatment.

Nowhere was this growing trend better expressed than in a 1941 lecture by Robert T. Frank at the New York Academy of Medicine. Frank was an early twentieth-century gynecologist who worked on ovarian endocrinology and anticipated the work of Allen and Doisy, but he remained a figure of minor importance because he was never able to bring his findings into sharp focus. Nevertheless, his lecture was both typical and revealing. "The menopause," Frank told his listeners, "is anticipated with dread by a large number of women, because they expect to lose their attractiveness (with appearance of wrinkles, gray hair, flabbiness, hirsutism), sex allure, physical vigor, and shapeliness (obesity, angularity). They likewise anticipate a diminution of libido and of mental capacity." The prevailing opinion at that time was that neurovascular, digestive, arthritic, and psychiatric symptoms were common in menopause. But, said Frank, all was not lost. "Those of us who have been interested in endocrinology for many years consider the estrogenic relief of the menopause as a major triumph, second only to that of the treatment of hypothyroidism by thyroid medication and of diabetes by insulin."[11]

During the 1920s and 1930s, endocrinologists and physiologists began to explore a variety of physiological processes related to the reproductive cycle. Some of their findings had significant implications for understanding and influencing bone development. Ultimately they transformed the understanding of the human skeleton, with the traditional concept of skeletal rigidity replaced by the idea of skeletal plasticity. Their research began to illuminate the process of demineralization and its relationship to organs (such as the parathyroid glands), as well as to calcium and phosphorus metabolisms; the role of other endocrine and metabolic disorders; and the effects of diets deficient in calcium, phosphorus, and vitamin D. Bones, in other words, were constantly undergoing change. During the first four decades of life, bones strengthened; in later life they lost mass, a process that was more rapid in women than in men. Osteoblasts were responsible for bone formation, whereas osteoclasts removed the mineralized matrix (they were responsible for bone resorption). Bone atrophy could thus have a variety of causes: disuse during immobilization, starvation and malnutrition, hypothyroidism, and excessive osteoclast activity.

In the mid-1920s the availability of parathyroid extracts allowed investigators to demonstrate that the parathyroid hormone affected the body's metabolism of calcium and phosphate. In 1929, young Fuller Albright (who was just launching a career that would make him one of the nation's most important clinical endocrinologists) and Read Ellsworth presented a complex explanation of the mode of action of the parathyroid hormone on bone. Subsequent research did not confirm their hypothesis, but those later findings did demonstrate that the parathyroid hormone had a decalcifying effect on bone.[12]

While some endocrinologists were studying bone development, others had become preoccupied with the reproductive cycle of higher forms of life. Was this cycle regulated by the nervous system or by the endocrine glands? When it became clear that the endocrine system played a crucial role, investigators began to shift their focus. Animal experiments, in particular, began to illuminate the relationship between ovarian function and calcium metabolism. In 1926 Oscar Riddle and Warren H. Reinhart (two biologists at the Carnegie Station for Experimental Evolution at Cold Spring Harbor, New York) detected a sharp increase in blood calcium among female pigeons during each ovulation period, a change not found in males. Inquiries were therefore needed to determine whether human reproductive periods—including menstruation, pregnancy, lactation, and menopause—were accompanied by changes in blood calcium levels. "The search for such gross metabolic changes is more likely to succeed," Riddle and Reinhart wrote, "if the glands of internal secretion are made the objects of investigation, since it is becoming clear that these organs are peculiarly and very intimately concerned in the phenomena of reproduction and of sex."[13]

In 1934 Preston Keys and Truman S. Potter reported that structural changes occurred in the bone marrow of female pigeons that were not found in males. The bone marrow of females underwent osseous modifications during the maturation of the ovarian follicle. In other words, the changes paralleled the female reproductive cycle.[14] Two years later Riddle and Louis B. Dotti found that estrogen had "great power . . . to increase the serum calcium in normal, castrate, hypophysectomized, thyroidectomized pigeons," as well as in other animals. Estrogen, therefore, played a crucial role in bone physiology.[15] Subsequently, Carroll A. Pfeiffer and William U. Gardner (two anatomists at the Yale University School of Medicine) undertook investigations into the effect of estrogen on the skeletal system. Working with birds and rodents, they found that estrogen stimulated the production of osteoblasts, and that the administration of estrogen to male pigeons resulted in bone formation similar to that of ovulating female pigeons. When estrogen was given to mice, the results were much the same: marrow cavities were replaced by compact bones. In 1943 Pfeiffer and Gardner concluded that their work on steroid hormones could possibly find "a practical application" in preventing or alleviating the symptoms of senile osteoporosis, or in accelerating the healing of fractures.[16]

Interest in bone physiology and its relationship to the endocrine system was obviously mounting in the 1920s and 1930s.[17] It is equally clear, however, that this interest had not yet created a diagnostic category. An important work by Henry A. Harris (a well-known English anatomist) that dealt with bone growth in healthy and diseased persons only briefly alluded to the association of adenoma (a benign tumor) of the parathyroid glands with osteitis fibrosa (a complication of hyperparathyroidism in which bones become soft and deformed). Harris's bibliography, moreover, did not mention the publications of Albright, Riddle, and others who had written about the relationship between bones and the endocrine glands.[18]

Studies dealing with the relationship between endocrine disorders and bone metabolism began to arouse interest in skeletal x-ray examinations, if only because biopsies were not possible. According to Ernst Lachman (a physician who had migrated from Germany and was employed at the University of Oklahoma Medical School) clinical experience demonstrated that radiography could diagnose many diseases, including parathyroidism, rickets, scurvy, and osteomalacia. Nevertheless, the x-ray method could not reveal whether bone atrophy was produced by halisteresis (the loss of salts, especially lime, from the bone) or by cellular resorption. Moreover, in order to become visible, the calcium loss had to be in the vicinity of 30–40 percent. It was clear that Lachman was defining osteoporosis in terms of bone atrophy that had a variety of different causes; it was not a discrete diagnostic category. X-rays, in other words, had serious limitations.[19] In later decades more sophisticated methods of imaging would contribute to the emergence of osteoporosis as a distinct disease or disorder.

FULLER ALBRIGHT AND
POSTMENOPAUSAL OSTEOPOROSIS

By 1940 it had become clear that bone development and the endocrine system were intimately related. Yet osteoporosis had not yet been elevated into a discrete category. It was precisely at this time that Fuller Albright introduced the diagnosis of postmenopausal osteoporosis.

Fuller Albright was born in 1900. He entered Harvard College at the age of 17 and graduated three years later. He then enrolled at the Harvard Medical School, where his encounters with patients began to shape his long-range interests. His natural curiosity was stimulated when he learned about the dramatic discovery of insulin. After receiving his M.D., Albright interned in medicine at the Massachusetts General Hospital and then spent a year conducting research with Joseph C. Aub. The latter's studies in lead poisoning stimulated Albright's interest in the metabolism of calcium. After a year as an assistant resident at the Johns Hopkins Hospital, Albright went to Vienna and studied under Jacob Erdheim, a renowned pathologist. Erdheim was the first to describe parathyroid hyperplasia in osteomalacia and the first to enunciate the calcioprotective law (on the topic of calcium deposits in bones), among many other contributions to endocrinology. His three years with Erdheim led to Albright's lifelong interest in bone disease, parathyroid function, and calcium metabolism. The rest of Albright's career was spent in research, teaching, and practice at the Massachusetts General Hospital.

Albright's research was always linked to the puzzles presented by his patients, rather than to the more abstract problems of biochemistry. He summarized his approach in his presidential address at the meeting of the American Society for Clinical Investigation in 1944. "I think of a clinical investigator as one trying to ride two horses—attempting to be an investigator and a clinician at one and the same time. . . . The rider of two horses, however, must remember that there are two horses; he must avoid the danger on one side that he, as a clinician, be swamped with patients and the equal danger on the other side that he, as an investigator, be segregated entirely from the bedside."[20]

In 1928 Albright and his collaborators undertook several studies (published the following year) dealing with calcium, phosphorus, bones, and the parathyroid glands. In 1937 he described a condition that became known as Albright's syndrome (precocious puberty in girls, cystic bone disease, and brownish skin pigmentation). Albright provided the first clinical description of hyperparathyroidism and called attention to its association with kidney stones. He also clarified the pathophysiology of Cushing's syndrome (a hormone disorder caused by high levels of cortisol in the blood). Albright was generally acknowledged as the preeminent clinical and investigative endocrinologist of his time. He developed Parkinson's disease in 1937, and in 1956 underwent an experimental surgical treatment that left him aphasic (sustaining damage to the part of the brain that controls language) and totally incapacitated. He spent the remaining thirteen years of his life attended by his wife and nurses at the Massachusetts General Hospital.

Albright was familiar with the work of biologists and endocrinologists who had illuminated the relationship of the ovary to the skeleton and had demonstrated that estrogen treatment increased bone mass in pigeons. His clinical experience with patients resulted in the publication of two papers (in 1940 and 1941) in which he defined what he called postmenopausal osteoporosis.[21] At that time fractures were classified in terms of the trauma that caused them. Albright's contribution was to redefine some fractures by relating them to physiological processes. He and his collaborators noted that estrogen created reserves of calcium in the trabeculae (tissue elements in the shape of a small supporting beam or rod), where they could be utilized during pregnancy and lactation by the fetus and newborn infant. The loss of estrogen following menopause caused bone loss. In other words, many women whose skeletons outlived their ovaries suffered bone loss. Fractures, in turn, followed the skeleton's inability to withstand normal biomechanical stress. Although skeletal fragility in older women had long been recognized, Albright was the first to associate it with menopause. Hence his diagnostic category of postmenopausal osteoporosis.[22]

Albright's study excluded all people over age 65. Bones, he pointed out, atrophy in old age. In a lecture before the American College of Physicians several years later, Albright amplified this observation. In older persons the skin was thin, muscle and bone mass decreased, the hair became scanty, the steroid-producing lands decreased their output, and certain secondary sexual characteristics were less prominent. Osteoporosis in old age was due to the loss of gonadal hormones (those from the ovary much earlier than ones from the testis). Hip fractures were a common clinical manifestation of the aging process. Osteoporosis among the elderly was thus designated as "senile osteoporosis."[23]

Of the forty-two patients in Albright's study, forty were postmenopausal women (having passed through normal menopause), ten of whom had experienced artificial menopause (surgical menopause brought on by removal of the ovaries). Albright had no explanation for the osteoporosis detected in the two men in his study, however. In no case did osteoporosis in the women occur before menopause. Albright therefore named this subcategory of osteoporosis "postmenopausal osteoporosis." It was an advanced, painful, and deforming condition that generally involved the spine and pelvis. Most of his subjects had severe disabilities that required them to use corsets, braces, crutches, and canes. The long bones were less likely to be involved, and the skull was almost never affected. Through clinical, chemical, and radiological examinations, Albright was able to exclude other conditions, such as osteomalacia and complications associated with Cushing's disease (a pituitary gland tumor, leading to the production of elevated levels of cortisol), Paget's disease (a chronic condition that can result in enlarged and misshapen bones), and hyperthyroidism. In osteomalacia or in rickets, calcium was not deposited in osteoid tissue, whereas in osteoporosis the osteoblasts were deficient in laying down osteoid tissue. The mechanism involved estrogen, which inhibited bone resorption.

Albright was primarily concerned with understanding bone physiology. Nevertheless, he and Edward C. Reifenstein Jr. (a former student who became a prominent endocrinologist) did not ignore possible therapeutic interventions for persons with severe disabilities. Since stresses and strains stimulated the activity of osteoblasts, physical activity, especially among older people, was important. Albright and Reifenstein were critical of orthopedic surgeons who immobilized their patients. The two researchers offered no recommendations about diet (excepting avoiding malnutrition), if only because its role in the etiology of osteoporosis was unknown. In cases of postmenopausal osteoporosis, which were often complicated by fractures, Albright and Reifenstein administered estrogens either alone or in combination with testosterone compounds. The results had been satisfactory, although there was a danger of sodium retention and a possible risk of cancer. To minimize the cancer risk, estrogenic therapy was suspended every four to six weeks, for one to two weeks at a time. Because Albright and Reifenstein thought that osteoporosis was not a disease involving the body's calcium and phosphorus metabolisms, high intakes of these minerals and vitamin D were not indicated (osteomalacia being the exception). The treatment of senile osteoporosis was the same as that for postmenopausal osteoporosis. Albright and Reifenstein believed that osteoporosis created severe disabilities that could be alleviated by medical intervention, but it was not a disease that affected large numbers of people.[24]

In a review of the records of approximately 200 postmenopausal women treated with estrogen therapy by Albright and his associates for one to twenty years, two physicians (from the Harvard Medical School and the Massachusetts General Hospital) reported that the progress of osteoporosis had been arrested in all instances, as judged by the measurement of an individual's total height and an x-ray examination of the spine. The incidence of carcinomas (malignant tumors) in the breast, cervix, and endometrium was low. Estrogen therapy, however, failed to demonstrate a single instance of an increase in bone density.[25]

The diagnosis of postmenopausal osteoporosis in the 1940s

and 1950s was quite limited. It generally involved women between the ages of 60 and 65 with *severe* symptoms. In a study of
234 women, one investigator provided a composite description of
their condition.

> The patient, usually a woman of about 60 years of age, first no-
> tices weakness and a dull ache in the lower part of the back.
> This may be aggravated, especially as the result of a slight jar
> or fall, to acute and often agonizing pain which may persist for
> weeks. As the condition progresses spinal deformity develops,
> usually a rounded kyphosis [hunchback] of the dorsal region,
> sometimes with reduction of the normal lordosis [inward curva-
> ture] of the lumbar region or else a compensatory exaggeration
> of this feature. There is usually considerable limitation of spinal
> movements due to muscle spasm. Tenderness on pressure over
> the spinous processes is not severe, but jarring or bending the
> spine usually elicits pain. The most striking radiological change
> is a uniform rarefaction of the affected bones, chiefly the spine
> and pelvis; the long bones and especially the skull remain com-
> paratively unaffected. In the lower dorsal and lumbar regions the
> vertebral bodies become biconcave and the intervertebral disks
> biconvex and thickened, the so-called "fish-spine" appearance.
> One or more of the biconcave vertebrae may be narrowed ante-
> riorly so that its body becomes wedge-shaped, while sometimes
> actual collapse of a vertebral body may be seen.

Treatment involved the administration of dienoestrol (a synthetic
estrogen), together with vitamin D and calcium glycerophos-
phate (to ensure high intakes of calcium and phosphorous). In
cases where vertebral bodies had collapsed, the use of some form
of spinal support was indicated, but prolonged immobilization
was to be avoided.[26]

One of the unforeseen consequences of Albright's work was
to strengthen the belief that osteoporosis was primarily a disorder that affected women. Albright himself rarely used gendered
language. Yet the emphasis on the role of menopause in bone loss
had the inadvertent effect of reinforcing the traditional belief that

women suffered more debilitating physiological and psychological problems than their male counterparts.

Albright did not think that his research had settled all of the problems of osteoporosis, as was evident in his concluding comments during his John Philips Memorial Lecture before the American College of Physicians in 1947. In modest words he noted:

1. I have told you more about osteoporosis than I know.
2. What I have told you is subject to change without notice.
3. I hope I have raised more questions than I have given answers.
4. In any case, as usual, a lot more work is necessary.[27]

BONE RESEARCH AND OSTEOPOROSIS

During the 1940s osteoporosis—as a disease—was not singled out as a major medical problem. Despite Albright's pioneering work, there were weaknesses in the knowledge base dealing with skeletal pathology and physiology. Equally significant, epidemiological data regarding the magnitude of fractures, especially among very old persons, was largely absent. There was considerable speculation, however, about the role of hormones, physical inactivity, diet, sex, age, menopause, the value of x-rays, and racial differences. Theoretical claims, generally unsupported by empirical data, were common.[28]

Albright's work had grown out of endocrinologists' interest in the role of sex hormones, a subject that was assuming an increasingly important role. Experiences with wound healing during World Wars I and II provided a further stimulus to research on bone physiology. These two conflicts hastened the transition from the idea of bone rigidity to the idea of its plasticity. Equally important was the realization that bone was a tissue that underwent long periods of growth and development, during which it was influenced by nutritional, hormonal, and environmental factors. Even when their growth was completed, bones were subjected to a variety of systemic disturbances.

Concern with wound healing also led to a series of twenty-two conferences, beginning in 1942 and ending in 1953, sponsored by the Josiah Macy Jr. Foundation and led by Albright. The sessions were multidisciplinary, serving as a bridge from a period of relative inactivity to a later era of intensified interest and research on calcified tissue. These conferences were followed by the Gordon Research Conferences, which began in 1954 and continued annually. The publication of the proceedings of these and other conferences dealing with calcified tissue during the 1950s and 1960s was indicative of the rising interest in bone physiology, as was the appearance of the journal *Calcified Tissue Research* in 1967 (which became *Calcified Tissue International* in 1979).[29]

Nevertheless, concepts of bone physiology and development during the 1950s and 1960s remained in a state of flux. "Osteoporosis," noted one physician in 1954, "has come to be a confusing subject because many who use the term do not have an understanding of its precise anatomical meaning." Was the decrease in bone density due to decreased osteoblastic activity, poor deposition of inorganic materials, or excess destruction? X-rays provided no answers to these important questions. Although osteoporosis was frequently seen in autopsies of older persons, it was impossible to determine its cause. "It would appear that there are numerous facets to the osteoporosis problem which have yet to be elucidated."[30]

By the 1960s there was a growing consensus that osteoporosis, however defined, constituted a serious health problem. Yet the evidence to confirm this consensus was extraordinarily weak. Many epidemiological studies were based on relatively small samples of patients diagnosed with osteoporosis. One study at the Henry Ford Hospital in Detroit involved 218 ambulatory women over 45 years of age. The investigators conceded that, with such a limited sample, it was impossible to determine the incidence of senile osteoporosis in the general population. They speculated that ethnic or racial factors were involved. Several other studies, based on very small populations, found a lower incidence of osteoporosis among African Americans, compared with whites.[31]

With the support of the World Health Organization (WHO), B. E. C. Nordin (a noted British physician who played a major

role in osteoporosis research during the second half of the twentieth century) conducted a preliminary survey of osteoporosis
among various countries in Africa, Asia, North America, Central
America, Europe, and Scandinavia in 1964. The starting point
was the WHO claim that in many parts of the world, people sub-
sisted on very low calcium intakes without suffering any ill effects to their bones. Based on x-ray findings, information on the
incidence of fractures among older persons, and discussions with
pathologists, orthopedic surgeons, and others, Nordin found that
the WHO hypothesis was not confirmed by his multinational
study. He concluded that both nutritional and aging factors were
involved in the development of osteoporosis.[32]

A year later a group of researchers from six nations who were
investigating mineral metabolism met to discuss the possibility of
creating an international collaborative study of osteoporosis and
fracture epidemiology. The participants in this initial meeting,
which was sponsored by the WHO and the National Institutes
of Health's (NIH) National Institute of Arthritis and Metabolic
Diseases, agreed that their inquiries should include three major
subjects: biochemical, radiologic, and dietary information; pathologic data; and data on fracture epidemiology. Limited funding
and the methodological problems of conducting a multinational
epidemiological study, however, probably became too difficult to
overcome, and the proposed study never came to fruition. In later
decades, however, the WHO would play an increasingly significant role in the osteoporosis community and in evolving public
policies on this topic.[33]

Still, it was obvious that there was little unanimity about bone
physiology, pathology, and development. To Reifenstein and others, osteoporosis was an endocrine disorder. The administration
of anabolic steroids (androgen and estrogen) to restore and/or
protect protein and osseous tissues in people with recognizable
senile osteoporosis, in older persons generally, and particularly in
women at and after menopause, minimized the risk of fractures.
Yet Reifenstein and Albright's hypothesis that a decreased synthe-
sis of bone matrix was the essential feature of osteoporosis had not

been substantiated by most isotope kinetic studies, and their theory tended to be refuted by histological evidence (minute tissue
structures discernible with a microscope) that suggested increased
bone resorption and normal bone formation.[34] Marshall R.
Urist (a prominent orthopedic surgeon and bone researcher at
the University of California, Los Angeles), also expressed reservations about Reifenstein's analysis: the majority of patients who
had various endocrine disorders did not develop osteoporosis; sex
hormone treatment was non-specific and without effect on the
cause of osteoporosis; and severe osteoporosis had been observed
in patients living on a low calcium diet, possibly leading to negative calcium balances for many years. Nordin believed that osteoporosis could be due to a long-continued negative calcium balance, which caused the removal of mineral from bone, followed
immediately by the removal of matrix. Thus osteoporosis in the
elderly was due to bone resorption. Urist, however, could find
no correlation between the dietary intake of calcium and osteoporosis. Instead he postulated an "antiosteoporosis" factor that
was present in patients who did not have osteoporosis, and was
deficient or absent in patients who did have it. This factor, Urist
suggested, may have been related to a person's ability to adapt to a
low-calcium diet. Like Reifenstein, Urist favored the administration of anabolic steroids, but he regarded them as empirical, in
that there was no rational explanation for their beneficial effects.[35]

Aside from different theories of bone physiology,[36] researchers
faced a basic diagnostic dilemma in the 1960s. Given that bone
loss was related to a universal aging process, when did this process
reach a point when it was designated as a disease? Despite circumstantial evidence, in 1963 Nordin noted at a symposium that
"the actual cause of primary osteoporosis is unclear." Though primary osteoporosis was extremely common in older persons, Nordin added that "there is every reason to believe that it represents
only a quantitative rather than a qualitative difference from the
normal state."[37] Urist and his associates distinguished between
what they called "physiologic" and "pathologic" osteoporosis. The
former was "the time-dependent, slow process of aging, atrophy,

or failure of retention of bone mass that occurs in proportion to reduction in muscle mass after age fifty." The latter involved "a disproportionate, rapid loss in bone mass, generally in the age interval between fifty-five and seventy years, and is characterized by *spontaneous collapse of vertebral bodies in the dorsal spine.*" This difference was crucial. In physiologic osteoporosis, fractures resulted from accidental falls or external forces; in the pathologic form, multiple fractures occurred spontaneously, without any significant external force.[38]

Was there a difference between physiological and pathological osteoporosis? The answer to this question raised fundamental issues. What was not clear, as G. Donald Whedon (Director of the National Institute of Arthritis, Metabolism, and Digestive Diseases from 1962 to 1981) observed at a symposium, was a definition of osteoporosis that, on the one hand, would deal with gradual bone loss as an accompaniment of aging, "and, on the other hand, with the fact that bone loss, whether normal or more rapid than normal, leads ultimately to a state which is incompatible with the terms 'health' or 'normality,' either through obvious structural breakdown (fracture) or in the likely risk thereof."[39] In a meta-analysis of thirty published studies, two investigators found that a decline in bone mass began at about age 35–45 in women, and around 45–65 in men. The rate of decline (as a percentage of the initial mean) was about 10 percent per decade in women and somewhat less in men. The average amount lost, however, was the same for both sexes. Every older person diagnosed as osteoporotic had less bone than younger individuals. But it had never been shown that the elderly people in these studies had less bone than other persons of the same age and sex who did not have osteoporosis. The amount of bone that a person had was a function of the amount of bone they began with, and of their age. "All people lose bone from the skeleton as they age," two clinical investigators noted. "There is little justification for continuing to regard senile or postmenopausal osteoporosis as a disease, distinct from the loss of bone with age." Moreover, among individuals with "thin bones," the incidence of fracture did not change with age.[40]

Paul D. Saville (a physician at New York City's Hospital for Special Surgery) demurred, however. "The incidence of osteoporosis and of fractures increases exponentially with age; so does that of cancer, heart-disease, and stroke. In fact, as mortals age the difference between disease and senescence gets less and less until eventually everyone is diseased; and the disease proves fatal."[41]

In 1960 Göran C. H. Bauer introduced the category of "osteopenia," by which he meant "too little calcified bone." A year later, he and his colleagues defined osteopenia as "any condition presenting less bone than normal. As used by us, osteopenia can develop at a low, normal, or high rate of mineralization, without any implication as to whether the primary abnormality lies in matrix formation, resorption, or mineralization."[42] The problem was that this term, which referred to the physiological aging process in bone, shed little light on diagnostic boundaries. It was obvious that there were variations in bone mass among people in the same age and sex categories. But why did some of those with thin bones suffer a fracture, while others in the same category did not? The diagnosis that bone was osteopenic or osteoporotic offered little guidance; it simply meant that the person had less bone that a healthy young adult in the prime of life, or less bone than a normal subject of the same age and sex. Moreover, the amount of bone that individuals had in their younger years varied significantly. In 1966 Nordin and his associates conceded that the pathogenesis (the origin and development) of osteoporosis was "obscure, although its association with certain endocrine disorders is well recognized. Its incidence is known to rise with age, but it is far from clear whether it is a 'normal' accompaniment of ageing." A year later Nordin raised a critical question: "When does this apparently universal ageing process deserve to be called a disease?" Normality, he observed,

> is defined in statistical terms and it is difficult to regard anything as abnormal which involves entire populations all over the world. It is however perfectly possible to suggest that there may be a normal and abnormal degree of this process and that the normal

degree should be defined on statistical grounds in the usual way. If "normal" and "abnormal" osteoporosis can be differentiated in this way it should be possible to detect in frequency distribution histograms a bimodal or at least a skewed distribution when the histograms are corrected for age. This is in fact the case. . . . [Nevertheless,] most of the patients called osteoporotic by orthopedic surgeons, radiologists, and physicians alike, are no different from their contemporaries who have been spared this label.

Nordin concluded that "one can only emphasize the state of prevailing ignorance about osteoporosis and hope that improved techniques will provide answers in the future to some of the many questions posed in this short paper."[43] Such discussions suggested that the diagnostic category of osteoporosis remained contested. Those who argued that it was a disease had to face opponents who believed that it was intrinsic to the aging process and therefore not necessarily pathological.

The problem was confounded because the classification of metabolic bone disorders in adults was by no means clear. Three conditions involved too little calcified bone: osteoporosis (too little formation of matrix); osteomalacia, or adult rickets (too little calcification of matrix); and osteitis fibrosa generalisata (too much resorption of matrix and calcium). Reifenstein was particularly concerned with senile osteoporosis, of which the postmenopausal form was a special type. Some degree of osteoporosis was present in all elderly people, but it could not be recognized because of the limitations of the prevailing diagnostic procedures.[44]

Radiographic techniques for diagnostic purposes had their own formidable problems. For indisputable x-ray evidence, the bone calcium content had to change by at least 30 percent or more, a process that could take from five to ten years.[45] Even when the loss of bone calcium was evident, the radiologist was not necessarily able to distinguish osteoporosis from osteomalacia.[46] At best, three London physicians conceded, "radiography only gives a rough guide to the incidence and the degree of osteoporosis."[47] Judging the degree of mineralization on routine x-rays was "hazardous," and techniques were "not consistently reproducible," noted two Lahey Clinic physicians.[48] To Lachman, several problems confronted the radiologist. What degree of demineralization had to be present before it could be diagnosed through radiology? Was the amount of decalcification necessary for a diagnosis the same in all bones, and was it the same in all parts of a single bone? What degrees of decalcification could be differentiated on the film? Moreover, he conceded, the boundaries between a normal state and an early pathological condition had not been clearly defined.[49]

The growing body of literature dealing with bone physiology and pathology was produced by a relatively small group of clinical investigators whose influence within the larger profession was limited. In 1947 Franklin C. McLean (a figure who made major contributions to the physiology of calcium metabolism and bone) complained that no comprehensive monograph or treatise on the physiology of bone had been produced. Even worse, textbooks dealing with this topic ignored the fact that bone had a physiology.[50] In the 1947 and 1953 editions of their book on bone biology, Joseph P. Weinmann and Harry Sicher devoted only a few pages to the healing of non-traumatic fractures. Their primary focus was on the structure, growth, and pathology of bones, and they ignored any discussion of therapies that might reverse bone atrophy.[51]

Although interest in bone physiology was intensifying, the results were less than gratifying. Research on connective tissue had yielded considerable information on collagen (the main component of connective tissue). Research on bone lagged, because of the technical difficulties introduced by calcified tissue and the poor solubility of bone collagen. "The crystal structure and chemical composition of the bone mineral," McLean and Urist wrote in the 1961 second edition of their important monograph *Bone*, "have been under investigation for more than one hundred years, and they remain a subject of intensive study." Yet the subject was still "in a fluid state" and dissenting views were common. Moreover, knowledge about resorption had not advanced since it was first identified by Kölliker in 1873. Kölliker had described resorption and named the osteoclast. Nearly a century later, researchers had added little to his speculations about the mechanisms that drove this process. Osteoporosis in adult life, McLean and Urist noted, was "refractory to treatment and irreversible with respect to recovery of normal density of the bones." Progress in this area would depend on knowledge about etiology. Until then, high-calcium diets or treatments with sex hormones would continue to be of "limited value."[52]

In his 1967 guest editorial for the first issue of *Calcified Tissue Research*, McLean was equally pessimistic about the then-current state of research. "The literature on the bone mineral, including the dynamics of its formation, [is] in a very unsatisfactory state; at the moment it cannot be said that the situation has improved substantially in the past 5 or 10 years. . . . There are numerous other areas of uncertainty, and even of controversy."[53] "Confusion exists in this field," echoed Harold M. Frost, an influential but often overlooked figure. "It is difficult," he added, "to extract sense from many texts and articles."[54] Such observations also held true in the succeeding decade. For example, in the 1975 fourth edition of their book on orthopedic diseases, Ernest Aegerter and John A. Kirkpatrick conceded that the elements that maintained normal osteoid balance were "complex and poorly understood." The divisions among medical specialties compounded the situation. Internists used the term "osteoporosis" to denote a clinical entity resulting from a variety of pathogenic mechanisms; orthopedists, to mean an area of weakness in bone; radiologists, to describe a localized region of abnormal radiolucency (allowing the passage of radiation, i.e., not very visible on x-rays); and pathologists, to mean a subnormal rate of osteoid production. None of these specialists understood what the others were talking about. Some attacked accepted principles by which metabolic processes were understood, but without offering alternative explanations, "a type of intellectual vandalism that creates more chaos than comprehension."[55]

The undeveloped state of knowledge regarding skeletal pathology and physiology, according to Frost, permitted numerous claims about pathogenesis to persist. Researchers offered a variety of theories about the etiology of osteoporosis: decreased calcium absorption, decreased estrogenic stimulus to osteoblastic activity, decreased dietary protein, an excess of adrenal corticosteroid hormones, diminished androgen secretion, a patient's genetic endowment, and a lack of physical exercise. Researchers also adhered to a simplistic theory that osteoblasts and osteoclasts, which were separate cell lines, acted independently of each other, and that these alone accounted for the skeleton's macroresponses to various challenges. Before the 1960s, the failure to recognize that osteoporoses were a diverse group of diseases inhibited research on them. As Frost observed as late as 1981, while the extraskeletal determinants of some osteoporoses were known, the causes of the most common ones (notably the postmenopausal and senile forms) remained obscure. The variety of theories inhibited a more sophisticated understanding of the osteoporoses and thus precluded the development of effective therapies.[56]

Frost contributed significantly to knowledge about osteoporosis in the latter half of the twentieth century. Like Albright, he generated many of his ideas from his role as a clinician working with patients. Nevertheless, his seminal ideas were often ridiculed or ignored by others. His early works appeared in such publications as the *Henry Ford Hospital Medical Bulletin*, because refereed journals refused to publish them. He conducted much of his research in his basement, using his own funds. In the 1960s he advanced a theory that the bone multicellular unit (BMU) was the key effector of bone metabolism. He insisted that osteoclasts and osteoblasts were involved in a coordinated bone remodeling sequence, and that bone resorption and bone formation were coupled. In normal bone, the number of BMUs, bone resorption rate, and bone formation rate were all relatively constant. Changes in any of these altered this steady-state situation. In adults, bone removed by a BMU was never completely restored; the result was a loss of bone with age. Although initially ignored, the BMU concept became a fundamental tenet in endocrinology. In 1965, Frost (in collaboration with Robert Hattner and Bruce N. Epker) published a landmark article in *Nature*, showing that more than 96 percent of bone formation in adults occurred only after resorptive processes. Frost also demonstrated that estrogen could preserve bone mass while decreasing bone formation, and he was a pioneer in the development of histomorphometry (a method that accurately quantified the level of cellular activity and the amount of existing bone mass), which was initially regarded as trivial and untrustworthy, but subsequently became a standard investigative technique. Many of Frost's concepts were sound, but it took decades for them to achieve respectability.[57]

Research on the etiology and physiology of osteoporosis was accelerating during the 1960s. Though many issues remained unresolved, this early, tenuous research was slowly beginning to impact clinical practice. During the 1960s the number of articles dealing with both research and clinical practice began to rise. A search of PubMed (the U.S. National Library of Medicine, National Institutes of Health's online biomedical literature database) under the heading "osteoporosis (limited to English)" revealed that during the 1940s only 34 articles were published; in the 1950s the number rose to 180; in the 1960s there was a dramatic increase, to 846 articles, many of which were intended to guide clinicians.

Many of the articles that provided advice to clinicians in the 1960s reflected the ambivalence that was characteristic in the bone research community. Involutional osteoporosis (age-related bone shrinkage) was due neither to a calcium deficiency nor the aging progress, insisted three University of Toronto physicians, but rather to decreased production of ovarian hormones. "If our premise that estrogen deficiency is the main cause of progressive bone mineral loss in aging females is correct, then ideally all estrogen-deficient women should be treated with a maintenance dose of that hormone."[58] Menopause, in other words, was a deficiency disorder, rather than a part of the life cycle. A group at the Mayo Clinic did not necessarily disagree that menopause was a contributing factor in the pathogenesis of osteoporosis, but they noted that the favorable short-term response to sex hormones in their study did not mean that bone resorption could be inhibited indefinitely by sex-hormone treatment, nor did it exclude the importance of other factors influencing bone resorption.[59]

To some researchers, osteoporosis was a nutritional disorder. In certain individuals, the dietary intake of calcium was low for most of their lives. For other people, their intake had been adequate, but their gastrointestinal mechanisms had prevented them from absorbing calcium. In both cases a gradual decrease in gonadal hormones further affected their calcium balance by increasing their urinary calcium loss, thus affecting bone formation. Increasing the intake of calcium, together with vitamin D, could thus inhibit bone resorption.[60]

Idiopathic osteoporosis (peculiar to certain individuals) presented the most puzzling problem to clinicians. It occurred in young adults who were otherwise well. Its clinical manifestations were similar to postmenopausal or senile osteoporosis, but those affected by it were neither postmenopausal nor senile. The cause of premature physiologic bone aging, however, remained a mystery. According to two University of Minnesota Medical Center internists, "many different stresses to bone metabolism may aggravate a tendency toward premature and excessive osteoporosis."[61]

A small minority—influenced by research on the beneficial impact of fluoridation in reducing teeth caries (cavities)—endorsed the use of sodium fluoride in treating osteoporosis, although the general consensus was that this therapy remained untested. Virtually all agreed that a sedentary lifestyle promoted the development of osteoporosis, and therefore they were supportive of physical activity. In general, therapeutic recommendations were eclectic.[62] In an editorial in the *Journal of Chronic Diseases*, Michael T. Harrison summed up the confusion characteristic of the 1960s. "The riddle of osteoporosis is unanswered," he observed. "There is evidence that deficiency of the sex hormones and of calcium both play a part, but the importance of each is unknown. Treatment possibly arrests the progress of the disease in some patients, but we are unable at present to cure it."[63]

The rising concern with bone disorders in general, and osteoporosis in particular, did not occur in a social vacuum. By the 1960s there was a heightened awareness of the growing numbers of older persons in the American population, and of the physiological changes and disabilities that accompanied this trend. According to one Mayo Clinic physician, elderly people were especially prone to falling and incurring a major fracture. Although making up less than 10 percent of the population, this segment accounted for more than 25 percent of all accidental deaths and 20 percent of all serious disabling injuries. The most common cause of injury in the aged was falling, and fractures were exacerbated by osteorporotic bones.[64] In a discussion of the aging process and disease, Manual Rodstein (a cardiologist and an associate medical director of the Diagnostic Services Division of the Equitable Life Assurance Society) detailed the changes linked with aging: a slowing of the division, growth, and repair of body cells; a decrease in one's recuperative ability after an injury or disease; impaired hearing and eyesight; a decrease in tissue elasticity; a decrease in the secretion of glands (leading to osteoporosis in females); a diminished ability to respond to stress; a decreased tolerance for drugs; and multiple diseases. "Those concerned with the care of the aged—family, physician, nurse, social worker, and administrator—must be made aware of these changes so that they may provide intelligent, gentle, patient, understanding, and compassionate care to the aged."[65]

Slowly but surely, osteoporosis was redefined primarily as a problem involving aged persons, largely females. "The great extent to which the human skeleton involutes with age," two endocrinologists wrote in an article in the prestigious journal *Science*, "is apparent in the high incidence of spontaneous vertebral and femoral fractures in elderly women."[66] The diagnosis of postmenopausal osteoporosis began to be merged into the category of senile and involutional osteoporosis. Statistical data were employed to confirm the dangers presented by bone loss. The claim was that ten million women had lost 30–40 percent of their vertebral bone; three to four million were known to have spinal osteoporosis; and 25 percent of all postmenopausal women had osteoporosis.[67] The greater incidence of senile osteoporosis was

attributed in part to humans' lengthening life span.[68] The fact
that there were few systematic epidemiological studies to support
such claims was all but ignored; the data were accepted at face
value. Nor were the diagnostic boundaries of osteoporosis clearly
described.

To define osteoporosis as a disease or condition of women reflected the traditional belief that women's characters were shaped
by their ovaries, and that in a postmenopausal state women suffered from a variety of physiological and psychological ills. In
subsequent decades the second-wave feminist movement (the first
being the suffrage movement that secured passage of the Nineteenth Amendment in 1919, giving women the right to vote)
would challenge both a male-dominated medical profession and
the view that menopause was a deficiency disease that led to decline. Interestingly enough, the gendering of osteoporosis stands
in contrast to fibromyalgia, a diagnostic category that lacked any
biopathology (biological cause). The prevalence of this diagnostic
category in clinical settings found that women constituted between 75 and 88 percent of the total number of people affected
by fibromyalgia. Yet virtually none of the research attempted to
explain the association between fibromyalgia and women, either
in terms of biology (sex) or in terms of culture (gender). Rheuma-
tologists avoided the temptation to feminize fibromyalgia, which
they defined as a musculoskeletal disorder with multiple tender
points. Using non-gendered biomedical language enabled them
to avoid a psychosomatic interpretation that would have stirred
the wrath of feminist scholars and made these clinicians susceptible to charges that they were once again suggesting women's in-
nate biological inferiority.[69]

In succeeding decades the diagnosis and boundaries of osteoporosis would be transformed. A combination of technological innovation (in the form of novel imaging machines), a redefinition of
menopause as a deficiency disorder, a faith that medical intervention—notably the use of hormone replacement therapy (HRT)—
could prevent osteoporosis and preserve femininity, pharmaceutical marketing, the creation of a group of endocrinologists and

orthopedists who made osteoporosis their sole focus, the internationalization of the diagnosis, widening diagnostic boundaries, and the creation of a nonprofit foundation to promote awareness of the condition served to elevate the importance of osteoporosis and transform it into a major public health problem.

CHAPTER THREE

The Transformation of Osteoporosis

Concerns with the diagnostic category of osteoporosis grew, and even accelerated, as the decades passed, which was reflected in the number of published articles on this topic. Articles listed under the heading "osteoporosis (limited to English)" in PubMed in the 1960s numbered 846. A decade later they had risen to 1,570, and during the 1980s the quantity again doubled, to 3,119. In the 1990s no fewer than 9,509 articles appeared, with a leap to 22,387 in the first decade of the twenty-first century. To Fuller Albright, postmenopausal osteoporosis was a condition that affected a very small group of women. During and after the 1970s, the boundaries of osteoporosis underwent a dramatic transformation and began to include millions of women.

Elevating osteoporosis to the status of a major public health problem was a process driven less by new medical discoveries than by a variety of social, cultural, and economic factors. By the late twentieth century the development of antibiotics and other drugs, the rapid expansion of biomedical knowledge, and the introduction of new technologies all combined to reshape both the theory and practice of medicine. An impression of irresistible progress was created. Sickness and disability were no longer perceived as inevitable consequences of life; they now represented problems to be conquered by scientific and medical advances. The gospel of health was spread by print journalism and television. Both conveyed a simple and attractive message: disease could be conquered if sufficient resources were devoted to research and health care. Scientific discovery, according to members of President Harry S. Truman's Scientific Research Board, was "the basis for our progress against poverty and disease." "The challenge of our times," board members added, "is to advance as rapidly as possible the understanding of diseases that still resist the skills of science, to find new and better ways of dealing with diseases for which some therapies are known, and to put this new knowledge effectively to work."[1]

This rising faith in the redemptive authority of medicine occurred as the structure of the American population was undergoing profound demographic changes. The decline in infant and child mortality in the late nineteenth and early twentieth centuries meant that more and more people survived into later life. Between 1900 and 1995, the number of Americans age 60 and over increased from 6 to 17 percent of the population. In 1995, over 43 million Americans (out of a total population of 262.8 million) were 60 or older, and 3.8 million were in the "very old" category (currently defined as those over 85). Before 1900, the mortality rate was the measure of health; thereafter, lifetime morbidity assumed an equal position. Chronic and long-term illnesses replaced infectious diseases as the major causes of death.

Such demographic changes had significant social, economic, and political implications. The aged began to coalesce into a significant bloc, with their own interests and agenda. They rejected the prevailing concept that aging and decrepitude went hand in hand. Robert N. Butler (a key figure in modern gerontology and first director of the National Institute on Aging from 1975 to 1982) referred to ageism as simply another form of bigotry. Ageism, he wrote in 1969, "reflects a deep-seated uneasiness on the part of the young and middle-aged—a personal revulsion to and distaste for growing old, disease, disability, and fear of powerlessness, 'uselessness,' and death."[2] Butler rejected the view that senility followed aging and instead attributed it to disease.

The traditional concept that aging was synonymous with decline and a loss of function was slowly but surely modified by a belief that aging and disease, though obviously related, involved two distinct processes. Although aging was inevitable, many aspects of aging could be addressed in ways that minimized a loss of function and expanded healthy years. Cataract surgery, hip and joint replacements, and control of systolic hypertension were but some of the therapies that contributed to the maintenance of function and an increase in longevity.

The medical profession indisputably helped alleviate diseases and disabilities associated with aging. The traditional emphasis on treatment, however, was augmented by the belief that medicine also possessed the capacity to overcome allegedly undesirable traits and prevent disease. Novel interventions seemingly could improve brain functioning, increase stature, arrest aging, increase longevity, alleviate anxiety and depression, create desirable character traits, maintain high levels of sexual activity, and reshape bodies, to cite only a few. A coalition of scientists, clinicians, and pharmaceutical companies promoted a variety of interventions (including drugs and surgery) that would presumably enhance both the physical condition and well-being of people, even those who were healthy.[3]

A culture that placed a premium on youth also sought ways of arresting the aging process and dealing with the disabilities associated with chronic conditions among the aged. Procedures and drugs to eliminate wrinkles, stop hair loss, and reduce other cosmetic aspects of aging became common. The medicalization of many facets of life, and especially of aging, also had a powerful gender element: women were the objects of scrutiny more often than men. Menopause was a prime example. It had been medicalized and pathologized and then redefined as an estrogen-deficiency disease. The bones of women, therefore, were far more vulnerable than those of their male counterparts and required medical intervention. It was within this context that the definition of the diagnosis of osteoporosis began to broaden dramatically, thus strengthening the argument that screening and treatment were vital in preventing the disabilities associated with the disease.[4]

These developments, in turn, created significant entrepreneurial opportunities for the pharmaceutical industry. Not only did the industry begin to promote drug therapy, but it also labored assiduously to raise public awareness about presumably underdiagnosed and undertreated conditions and helped to frame exaggerated prevalence estimates. Further, it created disease alliances (composed of drug company staff, physicians, and consumer groups) to expand the market for new pharmaceutical products. The industry was especially effective in creating fears about the consequences of disease and thus drawing attention to their latest products. Direct-to-consumer advertising fueled sales by leading the public to believe that drugs could alleviate, if not cure, many diseases. Multiple pharmaceutical companies also used their vast financial resources to co-opt the medical establishment.[5]

During these decades medicine also became preoccupied with risk factors as explanatory elements in the origins and causes of disease. Epidemiological studies of the relationship between smoking and lung cancer by Austin Bradford Hill and Richard Doll, the Framingham Heart Study, and the work of Ancel Keys on the behavioral origins of coronary heat disease shifted attention to the vital importance of risk factors. Risk factors were often conceptualized as a disease, thus requiring treatment. The number of epidemiological studies grew rapidly. Most employed cohort (a group of individuals having a statistical factor in common) or observational analyses. By monitoring disease rates and people's lifestyles, epidemiologists believed that they could identify risk factors and thus develop (or target) preventive strategies. The methodological problem was that the risk factors they identified were, at best, associations, but these factors did not establish causation. Nonetheless, such studies often became the basis of public health recommendations about what individuals should or should not do to prevent disease.

TECHNOLOGY

In 1970 there were few indications that fundamental changes were about to transform the diagnosis of osteoporosis. The previous year an International Symposium on Osteoporosis convened at the Montefiore Hospital and Medical Center in New York City. The papers presented there reflected the less-than-satisfactory state of knowledge about the diagnosis. G. Alan Rose (a London physician) emphasized that osteoporosis was easy to define but very difficult to measure. "Different bones rarify at different rates," he observed, "and, furthermore, despite all the work so far we still lack knowledge of the mean and standard deviation of calcium content of each bone at each age." It was obvious, Rose added, that a grossly osteoporotic spine was pathological in a 30-year-old individual and clearly physiological at age 80. But was the same spinal condition pathological or physiological at age 50?[6] In summing up the material presented at the conference, G. Donald Whedon noted that Albright's view about hormonal etiology was no longer accepted. "Many factors, hormonal, nutritional, physical, and circulatory . . . influence the rate of bone loss or bone preservation." Treatment, Whedon added, "is and still must be very broad, and the reasons lie in the multiplicity of etiologic factors involved and in the inconclusive state of our knowledge as to the relative significance or importance of each factor in a particular case. . . . Much more research needs to be done."[7]

Ambiguous and conflicting data were by no means the only problem confronting clinicians and researchers. Equally significant was the lack of rigorous and objective criteria for diagnosing osteoporosis. There was no single specific test; the diagnosis was made by combined clinical, chemical, and radiological findings, none of which—singularly or in combination—could establish clear diagnostic boundaries. Between the 1940s and 1960s, a variety of techniques for measuring bone mineral density (BMD) were developed. Some required bone specimens, while others employed high levels of radiation; clearly, neither testing approach could or would be widely used.

By the late 1960s, and especially in succeeding decades, technology began to become more prominent. Some turned to the newly developed computed tomography techniques to assess the mineral status of the appendicular skeleton; others employed Compton gamma ray spectroscopy to determine the presence and degree of osteoporosis in patients; still others used radiographic densitometry, microradioscopy, microradiography, and total-body neutron-activation analysis.[8] A major innovation was the development of single photon absorptiometry (SPA), which measured the BMD of the forearm, in the 1960s. This technology became widely available in the 1970s. It was a non-invasive procedure and exposed the patient to minimal radiation. Its use could be delegated to specially trained paramedical personnel, thus permitting the screening of individuals who were suspected to be osteoporotic. By 1980 double photon absorptiometry (DPA), which measured the bones of the spine, had come into use. The number of firms manufacturing these devices also rose; in the mid-1980s five companies were manufacturing SPA machines and four, DPA devices. Technology that is readily available, of course, is more likely to be deployed. Consequently, the number of screening centers, either hospital affiliated or owned by physicians, grew.[9]

While useful, the new technologies presented uncertainties. In one study of a high-calcium and high-phosphorous regimen among normal and osteoporotic populations, the results suggested either an insensitivity in the SPA method or the ineffectiveness of diet. In another study comparing SPA and DPA, difficulties in correcting for differences in the soft-tissue covering of the spine led to variations in the baseline from scan to scan, and problems of defining edges led to false high values. With distal arm–scanning procedures, slight movements of the forearm reduced precision. Most importantly, it was difficult to interpret the findings of these imaging technologies. While it was necessary to distinguish between osteoporosis (porous bone) and osteopenia (poverty or deficiency of bone), there were, according to Stanley M. Garn and his colleagues, "no established standards for separating these two entities, since failure to form bone and loss of existing bone cannot be reliably differentiated from a single radiograph."[10] These researchers' beliefs were by no means unique. According to a group associated with the Burke Rehabilitation Center, it was to be hoped that the inadequate methods for the early diagnosis and preventive management of bone loss "can be clarified by means of quantitative data rather than subjective impressions, which have too long prevailed."[11] An early diagnosis of osteoporosis, observed Friedrich Kuhlencordt (a Hamburg clinician), was "utopian because of the wide range of variation of bone mass in healthy people of all age groups." This was as true for the histomorphometry (a study of the microscopic organization and structure of a tissue) of bone biopsies as it was for the numerous radiographic methods for determining mineral content. The situation, Kuhlencordt noted in somewhat pessimistic words, "cannot be improved by increasing the precision of measurement of the different methods."[12] The absence of diagnostic criteria prevented the separation of those with osteoporosis from normal persons of the same age.[13]

In medicine generally, imaging held out the hope of diagnosing many diseases, of gaining a fuller understanding of pathological processes, and of developing efficacious therapies. In an editorial dealing with papers presented at the Third International Conference on Bone Mineral Measurement in 1976, Joseph P. Whalen noted that an accurate measurement of bone mineral content was potentially "an important tool in the early clinical diagnosis of certain metabolic bone diseases, in particular, osteoporosis." Furthermore, such measurements held out hope of early detection and thus would lead to appropriate therapeutic interventions.[14]

The technological innovations in the 1970s did not immediately lead to significant breakthroughs in understanding the pathogenesis of osteoporosis. A note of pessimism—accompanied by the hope of future progress—ran through much of the medical literature during this decade. Most clinicians and researchers believed that the majority of cases of osteoporosis were inadequately explained and therefore were termed "idiopathic."[15]

To a minority, the very diagnostic category of osteoporosis raised serious questions. All persons, according to clinical investigators H. P. Newton-John and D. Brian Morgan, lost bone as they grew older; the amount lost was a function of the amount of bone a person had at maturity. Excessive loss occurred only in the presence of diseases such as Cushing's syndrome and affected relatively few persons. Hence it was the amount of bone—not the rate of loss—that determined morbidity. To put it another way, there was no evidence that osteoporosis was the result of an excessive or abnormal loss of bone. "Fractures," these clinicians insisted,

> occur in the persons with the thinner bones and we have accounted for the increased frequency of fractures with age on the basis of the loss of bone with age. This finding leads one to suggest that fracture is a random event among persons with thin bones, and that other factors in the etiology of fractures, such as local changes in bone tissue and the trauma, do not increase in frequency with age. . . . [Osteoporosis] is a description of a state and not a disease. The increased frequency of "osteoporosis" with age can be accounted for by the universal loss of bone with age.[16]

Surprisingly for us today, there was relatively little concern with hip fractures in the 1970s. To most clinicians at the time, the hallmark of osteoporosis—largely among the elderly and especially among women—was vertebral compression fractures, which were often asymptomatic and rarely produced morbidity. Fractures of the neck of the femur (thigh bone), the distal radius (wrist), the proximal humerus (shoulder), and the pelvis were the primary causes of what were seen as non-osteoporosis-related disabilities.[17]

THERAPEUTIC DEBATES

The disputed debates about the pathogenesis of osteoporosis were matched by equally contentious discussions over therapy. At that time randomized controlled trials (RCTs) to determine efficacy were not especially common; therapeutic recommendations grew out of the clinical experiences of individual practitioners.

In a book intended for clinicians, published in 1976, Gilbert S. Gordan and Cynthia Vaughan observed that while basic research continued to add to medical knowledge, "much of the current reporting is conflicting and does little to clarify the confusion for the practicing clinician who must, perforce, deal with clinical actualities. . . . Confusion of research with treatment has led to some untoward assumptions about osteoporosis and premature application of unproved and, in several cases, painful and toxic diagnostic and therapeutic methods." Few of the endpoints (reasons for cessation) of therapy stood the test of time. Bone loss in osteoporotic adults seemed to be irreversible. "Unfortunately," Gordan and Vaughan wrote, "many promising therapies still in the stage of clinical investigation have been widely adopted before their therapeutic worth has been established."[18]

The diversity of therapeutic recommendations was perhaps best revealed in discussions dealing with postmenopausal osteoporosis. To some, estrogen treatment was valuable to prevent or retard the loss of bone mass, and these investigators reported that there was no evidence to support the claim that the administration of estrogen increased the incidence of breast cancer. Several studies, as a matter of fact, concluded that estrogen delayed the onset of breast cancer (except among those with a family history of breast cancer or a previous breast malignancy). The aging process, which presumably began with the cessation of ovarian function, was neither physiological nor inevitable. The administration of estrogen had the ability to replace the natural inadequacy of aging female gonads. It would suppress menopausal symptoms, retard physical atrophic changes and the development of atherosclerosis (an accumulation of fatty materials on arterial walls), and prevent osteoporosis.[19]

The claims about the benefits of estrogen therapy did not go unchallenged. If menopause was a normal, albeit annoying, phase of life, therapy was illegitimate. Moreover, the revelation of risk factors in the estrogen component of birth control pills was an issue of concern. One study, for example, found an unexpectedly high number of endometrial carcinomas among women taking estrogen, a finding confirmed by others.[20] Did the risks associated with estrogen therapy outweigh any potential benefits? The absence of data from studies evaluating the effect of estrogen on osteoporosis, heart disease, and cancer, noted a Yale physician, "are among the most pressing and deserving areas of biomedical research." An article in *Consumer Reports* expressed skepticism about the benefits of estrogen therapy for osteoporosis, referring to its use as "an unresolved issue."[21]

Perhaps no figure was more critical about the state of research than Robert P. Heaney (of Creighton University's Osteoporosis Research Center). He condemned the lack of scholarliness in evaluating estrogen therapy and the widespread medical and lay enthusiasm for what he called "supposedly definitive findings— often with little scientific corroboration." "Investigators have frequently set out to prove hypotheses correct rather than to test them," Heaney noted. It was clear that estrogen could not reverse bone loss; estrogen was effective only if its use was confined to palliation. Perhaps this hormone might play a role in prevention, but there were a multiplicity of unanswered questions. "To whom should prevention be offered?" Heaney asked. "Should it be all postmenopausal women, when only about one fourth appear to be at risk? And who are the women in the risk group? . . . Finally, what of dosage? It is not surprising that, with the paucity of qualitative data, there are at present no quantitative data." The morbidity and mortality of osteoporosis justified incurring the small risk of endometrial carcinomas associated with estrogen therapy, but it could not be justified for patients who were never at risk.[22]

Nor was there any consensus about the value of a variety of other therapies. Disagreements about the use of calcium supplements (with or without vitamin D), calcitonin (a hormone that has the ability to decrease blood calcium levels and suppress bone resorption), fluorides, and high-protein diets were common.[23] Even the manner in which physicians dealt with osteoporosis was a source of controversy. Urial S. Barzel (a well-known endocrinologist at the Montefiore Medical Center in New York City) thought that too many doctors approached osteoporosis as a static disorder in which the only things to be considered were bone density, strength, and breakability. Given the fact that the skeleton was involved, especially the vertebrae, Barzel believed that such an approach was not reasonable.[24]

Most of the medical literature dealing with osteoporosis emphasized the importance of research that would illuminate its etiology, thus leading to the development of more effective interventions to prevent and treat the disorder. There was virtually no interest in dealing with accidents that led to fractures among the aged. A notable exception was Mary Wheeler (a practicing Brooklyn endocrinologist who was also associated with New York State's Downstate Medical Center), who pointed out that accidents and mortality among the aged were linked with fractures. "The two most hazardous areas for older patients are the streets and the bathroom. Patients who are unsteady on their feet should be provided with a cane and instructed never to leave the house without it. Patients with osteoporosis should never walk in snow or on ice. Bathtubs should be provided with railings or other solid support for grasping. All loose rugs should be removed, and the free edges of larger carpets nailed to the floor. . . . Heavy lifting should be cautioned against."[25]

Wheeler's advice, however, was rarely echoed in the medical press. Clinicians and researchers alike were preoccupied with the medical treatment of osteoporosis and did not consider the potential hazards of the environments in which their patients lived. A decade after the publication of Wheeler's article, however, a group of clinical epidemiologists pointed out that there were two general strategies to prevent osteoporotic fractures. The first was to find ways to strengthen bones. The second was to find ways of preventing falls that resulted in fractures. "It is feasible," they wrote, "to treat some of the medical conditions associated with falling; many of the environmental hazards that may contribute to falling can be modified; and it is theoretically possible that the neuromuscular dysfunction that contributes to falling could be delayed or improved by exercise or specific types of physical training."

Surprisingly, they added, little attention had been paid to identifying risk factors for falls and describing preventive measures.[26] Animal research on osteoporosis offered the possibility that certain substances could inhibit bone resorption. In 1969 Herbert Fleisch (head of the Department of Pathophysiology at the University of Berne in Switzerland) and his associates experimented with diphosphonates (later known as bisphosphonates) and found that this category of substances could prevent the development of immobilization osteoporosis in rats. Moreover, the authors determined that several diphosphonates reduced blood calcium values in normal and thyroparathyroidectomized (removal of the thyroid and parathyroid glands) rats fed a low-calcium diet. This suggested that diphosphonates had the ability to reduce bone resorption both in the presence or absence of parathyroid hormones. Shortly thereafter, Miguel E. Cabanela and Jenifer Jowsey (two Mayo Clinic physicians) confirmed some of Fleisch and his colleagues' findings, but Cabanela and Jowsey determined that it was incorrect to extrapolate from disuse osteoporosis to systemic bone-loss disorders. In 1972 two other researchers, Joseph Lane and Marvin Steinberg, agreed that diphosphonates could inhibit disuse osteoporosis in adult animals, calling it "a highly promising therapeutic approach toward mastering osteoporosis."[27]

At this point pharmaceutical companies manifested relatively little interest in developing drugs (other than estrogens) to treat osteoporosis. A clinical and research community concerned with osteoporosis was beginning to emerge, but osteoporosis had not yet caught the attention of a broad general public. The absence of a large consumer market probably inhibited the search for drugs to treat osteoporosis. By the 1990s, however, the situation was quite different. By then osteoporosis was identified as a major public health hazard that affected millions of Americans. Under these circumstances, the pharmaceutical industry turned its attention to osteoporosis, with, as we shall see in chapter 4, the first of the prescription bisphosphonates being developed and brought to market.

THE ESTROGEN REVOLUTION

While a relatively small group of physicians and medical scientists were engaged in discussions of osteoporosis during the 1960s and 1970s, a larger debate over menopause and femininity was taking place. In the two decades following the isolation of estrogen by Edgar Allen and Edward Doisy in 1929, this drug was prescribed for a relatively small number of women to alleviate some of the symptoms of menopause. The work of William H. Masters and his colleagues, however, began a process that would help reshape concepts of female aging.[28]

Born in 1915, Masters received his M.D. at the University of Rochester, where he met George Corner, a leader in the field of reproductive biology and the codiscoverer of progesterone. With Corner as his mentor, Masters began work on the estrous cycle of rabbits, which strengthened his interest in human sexuality. In the 1940s and 1950s, Masters began to investigate the relationship between sex hormones and aging. He and his colleagues administered sex hormones to elderly female residents at the City Infirmary in Saint Louis, an institution that cared for indigent, infirm, and aged persons. In their first publication (in 1948), the investigators reported that they had induced menstrual periods in a group of thirteen women age 64–82. Out of this came a belief that would guide Masters's future research program: giving sex hormones (estrogen and progesterone) to older women induced "a reversal of one of the inevitable changes of aging" and provided an opportunity to see whether other manifestations of aging could also be reversed. In other words, the "retrogressive changes" that occurred in the ovaries at the time of menopause were preventable.[29]

By 1951 Masters—despite equivocal findings—concluded that HRT had the potential to revolutionize gerontology and the ways in which physicians treated America's older women. Masters emphasized that increasing numbers of an indigent elderly population were placing a heavy burden on American society, to say nothing about the waste of manpower caused by reduced functionality that could potentially be reversed. His partial solution, whether for indigent or middle- and upper-class women, was clear: HRT would lead to the "rejuvenation of useful mental and physical function for the aged." Insofar as men were concerned, Masters noted that endocrine deficiency might explain male incapacitation, but he did not pursue this line of research.[30]

Masters undertook a campaign to persuade obstetricians, gynecologists, and gerontologists to consider the virtues of sex hormones. He even developed the concept of "the neutral gender" that included all persons who reached the age of 60. All such individuals were theoretically capable of functioning, but they were virtual castrates exhibiting physical and mental degeneration. HRT could prevent or reverse this deterioration. Although he had done practically no research on men, Masters supported unisex medication (estrogen and testosterone in a 1:20 ratio). HRT would not extend longevity, but it would enable older people to live a better-adjusted and more useful life, as well as provide preventive benefits.[31] In effect, Masters coupled the concepts of aging and decline with the belief that maintaining femininity meant reversing the consequences of menopause.

By 1958 Masters had turned his attention from HRT to the diagnosis and treatment of human sexual disorders. His original message, however, resonated among other physicians, especially K. Kost Shelton (a clinical professor of medicine at the University of California, Los Angeles, and director emeritus of the Endocrine Clinic at the Los Angeles General Hospital), who believed long-term HRT would not only prevent osteoporosis and coronary heart disease, but would also help maintain a youthful appearance, a happy disposition, a good marriage, femininity, and sexual allure. Both Shelton and Masters (who knew each other) insisted that physicians had an obligation to provide women with the benefits of modern medicine; to do otherwise was reactionary.[32]

Robert A. Wilson (a New York gynecologist) also played a significant role in bringing HRT to the public's attention. In 1963 he and his wife coauthored an article entitled "The Fate of the Non-treated Postmenopausal Woman: A Plea for the Maintenance of

Adequate Estrogen from Puberty to the Grave." In this piece they provided a gendered analysis that emphasized the negative aspects of menopause while completely rejecting the existence of a true male climacterium. "A man remains a man until the end," but the drastic decline in estrogen among postmenopausal women led to hypertension, osteoporosis, psychological problems (melancholia and "a vapid cow-like feeling called a 'negative state'"), and desexualization, to cite only a few of the devastating medical conditions listed by the Wilsons. Their descriptions of elderly women were dramatic. "Most elderly women in the past looked, felt, and acted old. Stiff, frail, bent, wrinkled, and apathetic, they stumbled through their remaining years. The amount and variety of suffering was great. . . . There is ample evidence that the course of history has been changed not only by the presence of estrogen, but by its absence. The untold misery of alcoholism, drug addiction, divorce, and broken homes caused by these unstable, estrogen-starved women cannot be presented in statistical form."[33]

Three years later Robert Wilson published *Feminine Forever*. He began by pointing to a "biological revolution" that would enable women to retain "their full physical and mental facilities" and overcome the major medical problems associated with menopause, which, to him, was synonymous with castration. "Male provincialism" had led the medical profession to ignore the physical, social, and psychological problems that followed menopause. Estrogen therapy was the magic elixir. The "emotional liberation of women ranks as one of the greatest contributions to human progress that this century has produced." Wilson recalled his mother's tragic decline during menopause, when she was transformed from a wonderful woman into a "pain-racked, petulant individual." At the same time, he complained of professional indifference to his achievement in calling attention to the redemptive role of estrogen, and noted that such figures as Edward Jenner (smallpox vaccine), Ignaz Semmelweiss (antiseptic procedures), Louis Pasteur (vaccination and pasteurization), and George Papanicolaou (pap smears) had also initially been rebuffed by medical authorities.[34]

Feminine Forever was a spectacular success, selling nearly 150,000 copies during its first year of publication. Wilson also created the Wilson Research Foundation to educate women about menopause and the need for estrogen therapy. The foundation received financial support from the pharmaceutical industry, which saw its sales of estrogen grow rapidly. Between 1966 and 1975, the number of estrogen prescriptions nearly doubled and the market value of non-contraceptive estrogen almost quadrupled.[35]

Wilson was by no means alone in emphasizing the pathological character of menopause and the need for estrogen therapy. In 1969 David R. Reuben published his best-selling *Everything You Always Wanted to Know about Sex, but were Afraid to Ask*. Menopause, Reuben wrote, was a defect in the evolution of human beings. His theory was that in the remote past, a person's average life expectancy was 30–35 years, and human organs were designed to last only that long. Modern medicine, however, transformed health and led to a dramatic increase in life spans. Organs such as the ovaries, however, ceased working at about age 40, even though women lived for thirty or more years after that. The diminution in the production of estrogen after menopause was catastrophic. Estrogen was the "entire basis for femininity"—physiological as well as psychological. Reuben's argument was that when this hormone is shut off, a woman "comes as close as she can to being a man." Aside from such physiological changes as coarsened features and obesity, women were no longer interested in sex and became "emotionally marooned as well." Nevertheless, the castration caused by Father Time could be reversed by replacing the lost hormone and thus assisting in the maintenance of femininity. Reuben's portrayal of evolution and his statements about the increase in life expectancy and the role of modern medicine bore little relationship to reality. Nevertheless, his book added to the chorus celebrating the miraculous qualities of estrogen.[36]

Supplemental estrogen had been available since the 1930s, but its use had always been limited. By the late 1960s and especially in the 1970s, however, popular and professional perceptions had undergone marked changes. Such widely distributed magazines as

Vogue, Good Housekeeping, Newsweek, Ladies' Home Journal, and others published articles that emphasized the benefits of estrogen. These pieces stated that menopause, while inevitable, was best managed under medical supervision. Drug companies—especially Ayerst Laboratories, whose drug Premarin dominated the field—actively promoted the positive effects of estrogen therapy. Their "informational" pamphlets were distributed to doctors' offices and thence to patients. The message was clear. One typical pamphlet noted that 90 percent of menopausal women experienced a variety of symptoms, including "physical and emotional turmoil, sagging and flabbiness of the breasts, tendency to put on weight, . . . embarrassingly copious perspiration, excruciating headaches, causeless crying spells, . . . senseless, addle-headed anxiety."[37]

More and more menopausal and postmenopausal women began taking estrogen and kept doing so for multiple years. Moreover, the introduction of oral contraceptives in 1960 meant that younger women, whether married or single, were also on long-term hormone therapy. Although some physicians expressed reservations about the extended of estrogen, their concerns were overwhelmed by the popularity of the drug. Nor was the vogue for estrogen therapy unique in terms of health care perceptions. During these years the marketing of so-called wonder drugs—antibiotics and tranquilizers—leaped exponentially as Americans became convinced that there was a remedy for every ailment. Caught up in the enthusiasm for drugs—which was exploited by an ever-growing pharmaceutical industry—patients came to demand that their physicians write prescriptions, and their physicians complied. The expansion of insurance coverage for both health care and prescriptions also made it less expensive to see physicians and purchase drugs.

In 1975, however, two studies published in the prestigious *New England Journal of Medicine* provided evidence that women who used estrogen were more likely to get endometrial cancer than non-users. Both were retrospective case-control studies that compared a group that had endometrial cancer with a control group that did not and looked at the proportion of women in each group using estrogen. The results showed an increase in this type of cancer among those taking estrogen. Retrospective studies (ones collecting information about the past history of participants to explain their present condition), on their own, do not provide conclusive evidence. Nevertheless, the two 1975 studies constituted a warning that long-term estrogen use might pose unacceptable risks to women's health. Their publication set off a contentious debate in the medical press and among professional organizations, indicating deep divisions over the use of estrogen.[38] In a commentary on both articles, Noel S. Weiss (a University of Washington epidemiologist) conceded that it would be prudent to infer that estrogens could be a cause of endometrial cancer, but he emphasized that many issues remained unresolved. To the question of whether estrogens should even be prescribed, existing data provided no answer. "Despite the urgent need for answers, there is little choice but to remain in the dark for a few years more."[39]

The debate about the risks and benefits of estrogen was not confined to the medical community. The U.S. Food and Drug Administration (FDA), which has jurisdiction over determining the safety and efficacy of drugs, and what constitutes adequate labeling (information for physicians and patients provided on drug labels), became involved in the debate over estrogen. In the early and mid-1970s, there was a heated controversy that involved the right of patients to have access to specific drug information (in the form of inserts in prescription drug packaging). The FDA generally relied on advisory committees to make recommendations about drugs. In 1977 the agency's Endocrinologic and Metabolic Drugs Advisory Committee began an assessment of the safety and efficacy of estrogen as a treatment for osteoporosis. Its recommendations reflected the same ambivalence characteristic of the larger medical community. Committee members agreed that estrogen prevented bone loss, but they felt that there was insufficient evidence to support the claim that this drug prevented fractures. The advisory committee supported estrogen therapy for two groups—women who had undergone hysterectomy and oophorectomy, and postmenopausal women with demonstrated radiologic evidence of bone loss or fractures—but its members were not willing to recommend estrogen therapy for all post-menopausal women.[40]

In 1979 the Medical Intelligence section of the *New England Journal of Medicine* published a review by Martin M. Quigley and Charles B. Hammond (both associated with Duke University's Division of Reproductive Endocrinology and the Department of Obstetrics and Gynecology). "Few controversies in the practice of medicine," they wrote, "have generated so much discussion as has the role of estrogen-replacement therapy in the climacteric women." Quigley and Hammond noted that medical literature was replete with articles emphasizing both the beneficial and the potentially detrimental effects of estrogen therapy. After review-ing all the pros and cons, these authors rejected the enthusiastic claims about estrogen that were so prevalent in the 1960s and early 1970s and agreed with those studies that posited a causal relation-ship between estrogen therapy and endometrial cancer. With the possible exception of its use as a prophylaxis against osteoporosis, the only specific indications for prescribing estrogen therapy were for women castrated before menopause and for those postmeno-pausal women with atrophic vaginal changes. The committee members conceded that estrogen retarded bone resorption, but noted that it would not increase bone density. Nor did the drug reduce the incidence of fractures when treatment was initiated in older postmenopausal women. Moreover, whenever estrogen was administered, it was to be used in the lowest available dose. The committee emphatically rejected the claim that estrogen would prevent aging.[41]

The debate over the benefits and risks of estrogen led the NIH to convene a Consensus Development Conference on Estrogen Use and Postmenopausal Women in September 1979. NIH began these types of conferences two years previously to produce con-sensus statements on important and controversial topics in medi-cine. These documents not only revealed the indeterminate and controversial state of many medical subjects, but also the growing role of the federal government in health issues. Conference attendees—biomedical research scientists, clinicians, and consumers—hoped to reach agreement on the safety and efficacy of the particular medical technology under discussion, including drugs, devices, and medical or surgical procedures.

In addition to the 1979 consensus conference recommendations on the use of estrogen for postmenopausal women, its participants devoted part of their deliberations to estrogen's role in dealing with osteoporosis. They concluded that estrogen therapy could retard bone loss if administered at the time of menopause, but it was unclear if slowing down bone loss would prevent the development of osteoporosis. Case-control studies that had not yet been published reported that estrogen use was associated with a decreased risk of osteoporosis-related fractures, although more data were required before its efficacy could be established. A curious inconsistency remained, though. In randomized trials, accelerated bone loss following the discontinuation of estrogen therapy removed any favorable effects of the treatment. But in case-control studies, the use of estrogen at any time in the past conferred some protection against fractures. The conference participants thus concluded that estrogen therapy represented "a promising approach to prevention of the widespread problem of hip fracture."[42]

The claim that those women using estrogen were at higher risk for endometrial cancer resulted in a stunning decline in its use. Between 1975 and 1980, the number of prescriptions for estrogen declined from about 28 to 14 million. Then that number began to rise. By 1985 it climbed to 20 million and reached 30 million in 1990. This reversal was due to new data that appeared to confirm the hypothesis that estrogen loss following menopause caused bone loss, which in turn led to osteoporosis. Prior to 1980, the relationship between bone loss and osteoporosis was considered likely but uncertain. Indeed, in 1972 the FDA decided to relabel the indications for estrogen; the drug was deemed "probably effective" in treating bone loss from postmenopausal osteoporosis.[43]

The relationship between estrogen and osteoporosis still remained problematic during that period. In 1979, however, a team of physicians led by Lila E. Nachtigall (all of whom were members of the Department of Obstetrics and Gynecology at the New York University Medical Center) published the results of a decade-long prospective study (one starting with the present condition of participants and following them into the future) dealing with the relationship between estrogen replacement therapy (ERT) and osteoporosis, breast and endometrial carcinomas, and cardiovascular and metabolic problems. Their study (referred to here as the "Nachtigall study") began in 1965 and included eighty-four pairs of women matched by age and disease, all of whom consented to participate in a double-blind prospective study. The participants were inpatients at the Goldwater Memorial Hospital in New York City (which opened in 1939 as a hospital for chronic diseases) and remained there for the duration of the study. A research nurse was given a list of the eighty-four matched pairs; she randomly selected which member of each pair would be assigned to the treatment group and which to the control group. The half chosen for treatment received daily doses of conjugated estrogen tablets and medroxyprogesterone acetate for seven days each month. (Progesterone helps prevent the thickening of the uterus, caused by estrogen, that brings on menstruation.) The investigators used doses of estrogen that were two to four times greater than the recommended dosage, in order to ensure that the study's findings were as clear as possible. The control group received a placebo that matched the appearance of the active medications. Once enrolled in the study, the women received annual physical examinations. Bone mass measurements were made by photon absorptiometry. The research physicians did not know which patients took the medications and which took the placebo. The code was broken thirteen times in the treated group and seventeen times in the control group, however, either because of a major medical complication or death.[44]

To assess the data at the end of the study, researchers divided the experimental and control groups into two subgroups. One subgroup was composed of women whose last menstrual period (LMP) occurred less than three years prior to the beginning of the study, and the second subgroup contained those women whose LMP was more than three years before the study's inception. Decreases in bone mass for the estrogen-treated groups were significantly less than for the placebo groups. Both placebo-treated subgroups lost bone mass during the ten-year period, while the estrogen-treated subgroups either gained bone mass (when treatment was begun within three years of one's LMP) or had minimal bone loss. There were seven fractures in the control group and none in the treated group. Nor was there an increase in the incidence of breast or endometrial cancer or mortality in the treated group. These dramatic results appeared to confirm the efficacy of estrogen in preventing osteoporosis. The authors concluded that the study demonstrated "that when instituted more than 3 years after the menopause, ERT virtually halts the osteoporotic process. When estrogens are therapeutically administered within 3 years of the menopause, the process can actually be reversed."[45]

Nachtigall's prospective study, despite the relatively small number of women included in it, offered contradictory results to those of researchers who had found a link between estrogen treatments and endometrial cancer. Yet important issues remained unaddressed. The conjugated estrogen used in the Nachtigall study included equilin (derived from the urine of pregnant horses), a substance that is foreign to the human body. It is unclear even at present if its effects are different from estrogens naturally produced in the body. Many of the studies of the relationship between cancer and HRT used different forms of estrogen, a fact that was generally ignored, despite its potential significance.[46]

The fact that the Nachtigall study had not been funded by the pharmaceutical industry strengthened its legitimacy. Ayerst Laboratories had originally agreed not only to supply Premarin and a placebo that were identical in appearance, but had also offered to provide financing for the entire study. Nachtigall's departmental chairperson, however, believed that such support would compromise the research, so he stated that New York University's

Department of Obstetrics and Gynecology would pay for the entire study, thus assuring its independence. The publication of the final results in 1979 seemed to provide definitive proof that estrogens, when used with progestins, were not only safe, but could prevent and arrest the progress of osteoporosis.[47]

The Nachtigall study was by no means unique. Other researchers were also investigating the potential benefits and risks of estrogen therapy. That same year a study at the Yale–New Haven Hospital in Connecticut attempted to evaluate the efficacy of estrogens in preventing what were common major limb fractures in postmenopausal women. It employed a retrospective case-control method to explore a beneficial rather than an adverse effect of treatment. Postmenopausal women admitted to the hospital with fractures of the hip or the distal radius were surveyed and matched with women admitted to the orthopedic service for other reasons. Eighty patients from each group were interviewed to learn if they had taken estrogens in any form, for any duration, and those who had were asked about the relationship of that ingestion to the onset of menopause. The data revealed that those taking estrogens were much less likely to suffer from hip or distal radius fractures.[48]

In the 1970s a Scottish group began to study the efficacy of estrogen therapy. In 1980 they published the results of their nine-year study on the prevention of spinal osteoporosis in one hundred women who had undergone an oophorectomy. The group found that a significant reduction in height occurred among the placebo-treated group, but not in the group treated with estrogen. These investigators were relatively cautious in their analysis and recommendations. They noted that bone behavior in oophorectomized women had never been shown to be qualitatively different from what occurred in postmenopausal women. But, in the absence of a technique to detect those women at risk for osteoporosis before they suffered significant bone loss, plus a lack of knowledge about the true incidence of osteoporosis, the Scottish researchers considered preventive estrogen therapy "inappropriate" as a treatment to be given to all postmenopausal women.

Their study did demonstrate, however, that low doses of estrogen prevented the progression of spinal osteoporosis.[49]

Nevertheless, the use of estrogen therapy remained contested in the early 1980s. Whedon wrote that therapeutic recommendations regarding estrogens were uncertain, "because of concern for their reported effects on blood coagulation and the recognized possibility that endometrial carcinoma may be stimulated by them."[50] An article in the *American Journal of Public Health* in 1980 dealt with the "epidemic" of endometrial cancer. Its authors linked the rise of this form of cancer, beginning in the mid-1960s and peaking in 1975, to ERT. The decline in estrogen sales that started in 1976 was matched by a concomitant decline in the incidence rates of endometrial cancer.[51] Beverly A. Mosher and Elizabeth M. Whelen (of the American Council on Science and Health, an organization founded in 1978 by a group of scientists who were concerned that many public policies relating to health and the environment lacked a sound scientific basis) also echoed the fear that the risks of estrogen treatments included endometrial cancer, thromboembolism (a blood clot blocking a blood vessel), coronary disease, and strokes, all of which appeared to be age and dose dependent. Estrogens belonged to a "category of pharmacological agents that must be used with extreme care."[52] Nevertheless, those who associated estrogen therapy with endometrial cancer did not go unchallenged. Two well-known Yale epidemiologists noted that means of detection and other biases in such studies merely indicated the need to develop rigorous scientific standards in the epidemiological methods used in case-control research.[53]

In a presentation at the White House Conference on Aging in late 1981, Lawrence G. Raisz (a leading figure in osteoporosis research) emphasized the large gaps in what was then known about osteoporosis. Many women who experienced postmenopausal bone loss did not develop clinical osteoporosis, and many of those treated with estrogens and calcium supplementation still developed fractures. Precise techniques for assessing the severity of the progression of osteoporosis and its response to treatment were lacking. Demonstrating the effectiveness of therapy required large clinical trials carried out for long periods of time, and the related high costs mandated that such trials be undertaken only if there was a reasonable likelihood that they would yield useful information.[54]

Many clinical researchers were ambivalent about estrogen therapy; they recognized that there were both risks and benefits, and their recommendations, therefore, tended to be somewhat equivocal. The Council on Scientific Affairs of the American Medical Association, for example, suggested that both the physician and the menopausal patient make a judgment based on "a reasonable assessment of the relative benefits and risks." "There is risk in using any drug," its members added, "and this must be balanced against the risk of not using it."[55] Yet balancing benefits and risks, especially when neither could be identified with any degree of precision, was extraordinarily difficult. Moreover, were all post-menopausal women candidates for estrogen therapy? Was it possible to distinguish rapid from normal bone loss? These and other issues confounded the osteoporosis research community. Yet responsibility for deciding whether to use estrogen was left in the hands of the postmenopausal patient and her physician, both of whom were expected to come to an informed course of action.[56]

Despite cautionary warnings, there were many who endorsed estrogen therapy without reservations. Epidemiological data revealed that the aging of the American population had led to a dramatic increase in the prevalence of fractures. By the early 1980s there were approximately 210,000 hip fractures. The consequences of this injury were severe, and included a three-week hospitalization period and a higher mortality rate than in persons of a similar age and gender who had not sustained a fracture. Those who survived often suffered permanent disability and dependency.[57] The costs of care and treatment for hip fractures were also substantial. Gilbert S. Gordan (of the University of California's San Francisco Medical Center) argued that these facts meant that prevention was an appropriate course of action. "The evidence is that most osteoporotic hip fractures can be prevented and the doses of estrogen necessary to prevent postmenopausal osteoporosis are small and safe. . . . If osteoporotic hip fractures, which constitute the bulk of the problem, can be prevented (and the evidence is that they can be) with a prophylaxis that is cheap and safe, then why not prevent them?"[58]

By the mid-1980s, attitudes toward ERT had shifted in its favor. Between 1975 and 1981, R. Don Gambrell Jr. and his colleagues followed thousands of postmenopausal women. They found that women who took estrogens and progestogens had a lower incidence of endometrial and breast cancers than women who either took estrogen alone or abstained altogether from hormone therapy. Moreover, estrogen therapy may have even afforded some protection against breast cancer. There was also evidence that estrogens could reduce the incidence of certain metabolic disorders, such as cardiovascular disease and osteoporosis. "Estrogen replacement," Gambrell concluded in 1986, "should no longer be withheld from postmenopausal women, because the benefits clearly exceed the risks, especially when the regimen includes progestogen."[59] These findings were confirmed by other studies.[60]

The Lipid Research Clinics Program's report on estrogen use and all-cause mortality in early 1983 also reported favorable outcomes. That study followed 2,269 white women, age 40–69, for 5.6 years. The risk of death among estrogen users, irrespective of their hysterectomy status, was 0.37 times that of non-users (3.4 per 1,000 vs. 9.3 per 1,000). This lower rate of mortality among estrogen users persisted even after multivariate adjustments were made for confounding factors. The investigators conceded that they could not explain the lower risk of death in estrogen users but claimed that they had provided evidence for an apparent protective effect.[61]

At an NIH Consensus Development Conference in April 1984, participants had clearly moved beyond the limited recommendations of the 1977 meeting. Their report emphasized the gravity of the problems posed by osteoporosis and claimed that the disease affected as many as 15 to 20 million persons. Nevertheless, diagnostic criteria were vague; the report referred to "low skeletal mass" without stipulating what this meant. In its findings, the report called attention to the fact that 1.3 million fractures attributable to the disease occurred annually among people age 45 and older. Among those who lived to be 90, 32 percent of the women and 17 percent of the men suffered a hip fracture. The cost of osteoporosis to the nation was about $3.8 billion annually. Its pathogenesis was complex, and included cellular, physiological, and metabolic factors. Among the many possible causes of osteoporosis, two stood out: estrogen deficiency and calcium deficiency. Consequently, the mainstays of its prevention and management were estrogen and calcium; exercise and nutrition might also be important adjuncts. Candidates for estrogen therapy included women whose ovaries had been removed before the age of 50. Those who had a natural menopause should be considered for ERT if they had no contraindications. Equally significant was the report's statement that the duration of estrogen therapy "need not be limited." Adding a progestogen to this treatment probably reduced the risk of endometrial cancer. The report emphasized the great need "for additional research on understanding the biology of human bone, defining persons at special risk, and developing safe, effective, and low-cost strategies for fracture prevention."[62]

That same month the FDA's Fertility and Maternal Health Drugs Advisory Committee met to review proposed revisions to estrogen package labels for both physicians and patients. Its members agreed that there was sufficient evidence to justify a recommendation that ERT should be given not only to postmenopausal patients with bone loss, but to *all* postmenopausal women as a preventive therapy. The committee ignored an earlier report given to its members that pointed out that there were few data to answer the question of how long preventive hormone therapy should be used. The committee members also recommended the addition of a progestin for seven days each month to prevent endometrial cancer. The following year the Endocrinologic and Metabolic Drugs Advisory Committee became involved in a semantic debate over the meaning of "treatment" as opposed to "prevention." A woman with osteoporosis received treatment, while prevention applied to someone with no evidence of osteoporosis but who was presumably at risk for developing the condition. Could estrogen be a "preventive treatment"? When the FDA issued its final labeling recommendation in 1986, it avoided the use of the term "prevention" while still implying that estrogen could slow bone loss.[63]

The debate over estrogen therapy had a major impact, irrespective of its merits. More than any other development, it had the effect of bringing the pervasiveness and dangers of osteoporosis to national attention. Prior to the 1980s, discussions of osteoporosis had been confined to a relatively small group of researchers and clinicians. Their preoccupation with its diagnosis, however, did not enter into the public's consciousness. The rationale for estrogen therapy, initially promoted as an antidote to the allegedly devastating consequences following the advent of menopause, expanded to include both the prevention of osteoporosis and its treatment. A survey by a public relations firm (commissioned by Ayerst Laboratories in 1982) found that more than three-quarters of American women had never even heard of osteoporosis. Both companies then engaged in an extensive public campaign to educate women and health care professionals about the dangers of osteoporosis and, not coincidentally, to promote the use of Premarin.[64] By the mid-1980s, the attention of a widespread general audience, which was overwhelmingly female, began to focus on the dangers of osteoporosis. It began to be common to refer to an "epidemic" of osteoporosis and to designate it as a major public health problem.

Cultural trends reinforced the growing preoccupation with osteoporosis. As historian Elizabeth S. Watkins pointed out, the decade of the 1980s emphasized youth and beauty. The accession of Ronald Reagan to the presidency embodied the virtues of youth. Despite the fact that he was the oldest person to occupy that office, Reagan—with his acting background, his dark hair, and his younger, attractive second wife—seemed to have overcome many of the undesirable traits associated with aging. Pharmaceutical companies implicitly promoted an anti-aging ideology by suggesting that their products could arrest the aging process and thus permit women (and men) to maintain a youthful, vigorous, attractive appearance.[65]

The publication of a series of popular books, some written by physicians, also endorsed an anti-aging mindset. The belief that individuals could potentially maintain health and vigor through their behavior and lifestyle choices reinforced the culture of youth. Yet a majority of these volumes also emphasized that medical science, especially for women, had an important role to play. Medical intervention (including drugs and a variety of surgical procedures) could overcome many of the undesirable attributes of aging. Moreover, medicine had the potential to mitigate multiple problems associated with menopause. Winnifred Berg Cutler (a biologist and founder of the Women's Midlife Wellness Center) and Celso-Ramón Garcia (a obstetrician-gynecologist with an international reputation in reproductive medicine), both at the University of Pennsylvania, published two volumes that provided an optimistic overview of the potential benefits in the medical management of menopause. The first book (with coauthor David A. Edwards) was designed to appeal to a broad public, and the second to physicians. The theme of both volumes was that the menopause required medical management. HRT could mitigate its symptoms, facilitate cardiovascular health, prevent or halt bone degeneration, positively affect natural immune responses, improve one's sense of well-being, and benefit facial skin.[66]

In a talk at an international congress dealing with menopause, Morris Notelovitz (a South African–trained obstetrician-gynecologist who was the founder and director of the Center for Climacteric Studies at the University of Florida) stated that "climacteric medicine is a discipline waiting to be born." The needs of women during the various stages of the climacteric required a physician's expertise, complemented by the skills and talents of nutritionists, exercise physiologists, psychologists, and social counselors. Climacteric medicine held out the promise of preventing or ameliorating chronic illnesses and thus avoiding the need for and cost of long-term geriatric care.[67]

Two years previously, in 1982, Notelovitz and Marsha Ware published *Stand Tall!: Every Woman's Guide to Preventing Osteoporosis*. This book emphasized the dangers of bone loss, which produced a variety of unattractive features associated with aging. Osteoporosis, according to Notelovitz and Ware, presented women with "one of the greatest health threats to your later years both in terms of the quality and the length of your life." Yet all was not lost. In collaboration with their individual physicians, women could design a lifestyle that would enhance their physical well-being, one that included a low-protein and high-calcium diet, exercise, and tobacco and alcohol avoidance. For some women, however, calcium and exercise alone were not sufficient; there were also women who refused to follow any regimen. For such individuals, hormone therapy was the "final option." The benefits of this therapy, the authors claimed, clearly outweighed the risks of endometrial cancer. Excerpts from *Stand Tall* appeared in numerous widely distributed periodicals and helped to popularize hormone therapy.[68]

Several other works by physicians avoided the fear tactics that pervaded *Stand Tall*. *The Osteoporotic Syndrome*, edited by Louis V. Avioli and published in 1983, included contributions by well-known clinicians and scientists and was designed to assist practicing physicians. The tone of most of the chapters was moderate and suggested that many issues remained unresolved. Robert Heaney, for example, emphasized that studies needed to identify risk factors had not been done, and that "the best one can do is to fall back on uncontrolled, largely anecdotal information." He noted that while older people had a propensity to fall, there was a striking absence of studies on postural and gait stability, which might explain why some people with decreased bone mass sustained fractures whereas others did not. The chapter authors in this first edition did not pay much attention to estrogen therapy. Although—as Avioli noted in the preface—osteoporosis was "preventable to a considerable degree," most of the contributors were relatively restrained in their discussions.[69] In a second edition, four years later, Robert Lindsay prepared a chapter devoted to estrogen therapy. As director of a bone research group, Lindsay found that administering ERT in postmenopausal women resulted in a maintenance of bone mass and, hence, a reduction in hip and femoral-neck fractures. Nevertheless, the inability to quantify risks for all patients meant that physicians would have to make their own judgments about initiating therapy on an individual basis.[70]

David F. Fardon's *Osteoporosis: Your Head Start on the Prevention and Treatment of Brittle Bones*, which appeared in 1985, also avoided enthusiastic claims and attempted to summarize the latest scientific findings about bone physiology. Fardon had no doubt that exercise was a partial solution, although there were many unknowns about its appropriate forms. Hormone treatment had protective effects, especially if taken during and immediately following menopause, but it was difficult to determine its overall risks and benefits. The decision to take estrogens required the collaboration of a knowledgeable physician. Proper nutrition was important, although good information on this topic was often obscured by "commercialism and the hype of enthusiasts." What was needed most was the identification and promotion of a lifestyle that would prevent the suffering that was associated with osteoporosis. Macmillan Publishers originally planned a major marketing effort to publicize the book, but for some unknown reason the firm reneged and the book sold fewer than 10,000 copies.[71]

In 1986 Lila Nachtigall collaborated with freelance journalist and writer Joan Rattner Heilman to produce *Estrogen: The Facts Can Change Your Life*. Nine years earlier the *Lila Nachtigall Report* was published, even before the results of her and her associates' randomized controlled trial on estrogen had become available. In the *Report*, Nachtigall dealt with the physiology of menopause. Her discussion of ERT was circumspect, and she presented both the pros and cons of the therapy.[72] In her subsequent book, however, Nachtigall wrote that ERT mitigated menopausal symptoms, preserved strong bones and thus prevented osteoporosis, helped maintain sexual activity in the years following menopause, minimized the risk of recurring urinary tract infections, and sustained smooth and younger-looking skin. Moreover, recent studies had demonstrated that estrogen-progesterone therapy was safe if employed correctly. Not all women required ERT, and those with a history of uterine cancer or estrogen-dependent breast cancer could not take the hormone safely, Nachtigall advised. But when menopause became a difficult phase, ERT was an indispensable adjunct. "I'll tell you what I tell my patients," Nachtigall wrote. "Fight back! Refuse to suffer needlessly. Don't accept major discomforts or disabilities when you don't have to. Take charge of your own body whenever you can. And make your own decisions—with the help, of course, of a competent and knowledgeable physician. . . . You can take the hormone systematically with pills or patches, or topically with vaginal cream. You can take it only as long as it helps you and stop taking it whenever you want to—or you can take it for the rest of your days without danger if you follow the rules."[73] *Estrogen: The Facts Can Change Your Life* sold roughly 100,000 copies. Updated editions appeared in 1995 and 2000, both of which also enjoyed widespread success.[74] Lila Nachtigall's *What Every Woman Should Know: Staying Healthy after 40* (cowritten with Robert D. Nachtigall and Joan Rattner Heilman) repeated many of these themes.[75]

During the 1980s the debate over estrogen treatments had a major, if unintended, consequence. Initially designed to mitigate the adverse consequences of menopause, estrogen therapy also enhanced the importance of the diagnostic category of osteoporosis and helped elevate its status to that of a significant public health hazard. In subsequent decades, as we shall see in chapter 6, the boundaries of the diagnosis continued its dramatic expansion. The stage was also set both for the development of a variety of other drugs that presumably could arrest the progress of bone loss, and for the emergence of a community of osteoporosis researchers that was international in scope.

CHAPTER FOUR

Popularizing a Diagnosis

By the mid-1980s, concern with osteoporosis was no longer confined to a small clinical and research community. Interest in the diagnosis had expanded dramatically in America and included not only a broad public composed of ever larger numbers of women, but also a variety of organizations representing different constituencies. Charles H. Chestnut III (an osteoporosis researcher at the University of Washington) recalled that when he gave presentations of his work in the 1970s, "there would be three or four people in the audience, and most of them were the next speakers." By 1995, research funding had increased dramatically and his presentations now drew hundreds of scientists.[1]

Yet fundamental disagreements within the medical and scientific community were common. There was no consensus over diagnostic boundaries, the role of risk factors, the importance of screening, the nature of bone loss, or even appropriate therapies. The weaknesses underlying the knowledge base about osteoporosis, however, were rarely transmitted to the general public. Warnings about the need to identify and treat a condition or a disease that threatened the health and well-being of the nation's women were the rule rather than the exception.

POPULARIZATION

Attention to osteoporosis was intensifying, and this was reflected in the creation of a national advocacy organization. In 1984 William A. Peck (a past president of the American Society of Bone and Mineral Research) convened a meeting of a group of leading researchers. As a result, the nonprofit Osteoporosis Foundation came into being, which was shortly thereafter renamed the National Osteoporosis Foundation (NOF). The foundation's objectives were to educate the public about and heighten awareness of osteoporosis, to educate physicians and other health providers, to develop strategies to gain funding for research and training, and to support other professional societies concerned with osteoporosis and metabolic bone disease. In a move intended to increase its influence in Congress, the foundation persuaded Representative Paul G. Rogers (a figure who played a major role in federal health legislation for two decades) to chair the NOF board.[2] The organization also maintained an intimate relationship with pharmaceutical firms, which contributed funds and provided grants to researchers (who also served as members of NOF's board of trustees).

In mid-1985 the Senate Subcommittee on Aging held a public hearing on osteoporosis, a development that also symbolized Americans' heightened interest in osteoporosis and the growing influence of a variety of groups dedicated to the promotion of women's health. Chaired by Senator Charles E. Grassley of Iowa, the committee included such prominent figures as Claiborne Pell of Rhode Island, Howard M. Metzenbaum of Ohio, and Paula Hawkins of Florida (who was responsible for persuading her colleagues to hold the hearing). In his opening remarks, Grassley noted that many people did not become aware of a decrease in their bone mass until they experienced a serious fracture. Perhaps 20 million women suffered from this condition, he stated, and 200,000 hip fractures per year were attributable to osteoporosis, to say nothing about vertebral collapses. The costs to the nation ranged from $4 to $7 billion annually. Yet osteoporosis was both preventable and treatable. Although research had been underway for several decades, noted the senator, the "public education effort . . . is just beginning." Grassley's remarks were echoed by Hawkins, Pell, and Metzenbaum. All referred to osteoporosis as a disease that afflicted older persons. Their comments clearly indicated that they had been briefed by prominent figures in the osteoporosis community and had accepted at face value what they had been told.[3]

Olympia J. Snowe, then a member of the House of Representatives from Maine, submitted a written statement in which she pointed to an increasing concern with osteoporosis. The previous year, Snowe had introduced a joint resolution designating the first week in May as National Osteoporosis Week. Although Congress took no action in 1984, the following year her resolution easily passed both houses of Congress and was signed into law by President Reagan on May 20, 1985.[4]

The first witness to testify at the hearing was Bernice Long (Tennessee's state coordinator for health advocacy), who worked with and represented the American Association of Retired Persons (AARP) in helping the public recognize the importance of osteoporosis. Despite having had annual medical checkups for many years, Long was not aware that she had developed a calcium deficiency. Her physicians never warned her that she was a potential candidate for osteoporosis. At the age of 72 she lifted a heavy package and fractured her spine. "I cannot describe to you the pain I experienced," she told the committee. Long wore a body brace for three months; luckily her fracture healed, and she had no long-term disability. Long believed that her calcium deficiency was responsible for her osteoporosis, and her system fortunately responded to "massive doses" of calcium supplements and a high-calcium diet.[5]

Long's faith in calcium supplementation reflected a belief that the calcium intake of a large sector of the general population was below the recommended daily allowance, which had a detrimental impact on health. An editorial in *Calcified Tissue International* suggested that calcium deficiency played a major role in the aging of humans. "If the international scientific community ultimately reviews the accumulated data regarding calcium intake and diseases such as osteoporosis, hypertension, senile dementia, etc., and conducts the appropriate experimentation," insisted its author, "perhaps a refreshing and more consistent theory of aging will evolve."[6]

The next two witnesses, T. Franklin Williams (director of the National Institute of Aging) and Mortimer B. Lipsett (director of the National Institute of Arthritis, Diabetes, and Digestive and Kidney Diseases), were far more cautious in their testimony. Williams emphasized that deficiencies in calcium and vitamin D were important etiological factors. Yet osteoporosis alone rarely caused hip fractures, which generally followed a fall. Research had demonstrated that certain individuals had "specific neuro-muscular deficits" that impaired their balance. Unfortunately, the pathophysiological mechanisms underlying these deficits were unknown. Williams pointed to gaps in knowledge about the effects of hormones, which restricted his ability to offer specific recommendations for preventive strategies. Much of the information on the efficacy of calcium supplementation, estrogen therapy, and exercise was derived from studies of younger persons; the effects of the regimens on older persons were unknown. Lipsett's testimony paralleled that of Williams. Lipsett emphasized the need for larger studies of risk factors and provided a list of grants that his institute supported. Lipsett also noted that calcium supplementation and estrogen replacement were factors in the prevention of osteoporosis, but he added that "the roles of other agents in one or another combination need to be worked out."[7]

Other medical authorities followed, including B. Lawrence Riggs (of the Mayo Clinic), Luella Klein (past president of the American College of Obstetricians and Gynecologists), G. Donald Whedon (retired director of the National Institute of Arthritis, Diabetes, and Digestive and Kidney Diseases), and William Peck (the first president of the newly formed NOF). They agreed that diets lacking calcium—particularly during adolescent years, when adult bone was being formed—had an influence. Among teenagers, soft drinks, which were rich in phosphates but low in calcium, had replaced dairy products, thereby inhibiting the formation of a denser skeleton. The diet of too many Americans, both young and old, was deficient in calcium and vitamin D. Decreased physical activity also played an etiological role. Estrogen therapy could retard bone loss, but there were serious risks involved. That existing treatments left much to be desired was clear. As Whedon noted, "treatment procedures are still unproven and have to be regarded as still in the research category."[8]

The claim that a diet lacking in calcium played an important role in the etiology of osteoporosis provided the dairy industry with an opportunity to further advertise its products in the same way that Ayerst Laboratories was promoting Premarin. Both industries launched national campaigns to inform the public about the health benefits of their wares. Gloria Kenney (vice president for nutrition of the National Dairy Council) offered testimony about the educational materials being disseminated to the public. Two years previously, Congress had created the National Dairy Promotion and Research Board to strengthen that industry's position in the marketplace and expand sales avenues for milk and dairy products. The Board was funded by a modest tax on milk and it used those monies to promote dairy products.[9]

The only critical statements came from the National Women's Health Network, a public-interest organization founded in 1975 and devoted exclusively to women's health. Since the 1960s, a resurgent feminist movement had begun to focus on issues pertaining to women's health and had become increasingly critical of a male-dominated medical profession. Their dissatisfaction was focused not only on the paternalistic and condescending manner of male physicians in treating their female patients, but also on such drastic and disfiguring therapies as radical mastectomies. Challenges to this procedure led to a dramatic decline in its use. Between 1974 and 1979, the number of radical mastectomies fell from 46,000 to 17,000, and within a few years less than 5 percent of breast-cancer operations would be radical mastectomies. This transformation was the result both of new findings that undermined the primacy of radical mastectomies, and a campaign in which women took a far more active role in issues pertaining to their health and well-being.[10]

In its statement at the 1985 public hearing, the National Women's Health Network commended the Senate Subcommittee on Aging for its concern with osteoporosis and noted the upsurge in public education through medical journals, women's magazines, advertisements, and newspaper health columns and articles. Nevertheless, this organization warned that there was "danger that existing treatment for osteoporosis may be mass marketed and that women will once again be guinea pigs for medical treatments which could prove to be fatal." Too many women had suffered ill effects and death as a result of taking prescribed hormones when further testing of these drugs was needed. Equally important, the vast majority of research projects were funded by pharmaceutical companies with a vested interest in magnifying the importance of osteoporosis. Other research projects were underwritten by food-product trade associations, which had their own agendas. The National Women's Health Network further stated that independent research also needed to include patterns of nutrition and exercise and not focus primarily on osteoporosis as a result of a hormonal deficiency. Because of sex-role stereotypes, a generation of the women who were studied had been groomed to be sedentary and slim, which in turn ruled out regular exercise and promoted poor eating habits and crash diets. The organization opposed hormonal treatment for all women, although it conceded that hormones could be prescribed in special circumstances if the women taking them were monitored for short- and long-term side effects.[11]

The subcommittee's deliberations reflected the consensus on the nature, prevention, and treatment of osteoporosis that prevailed in the mid-1980s. Osteoporosis posed a serious medical and economic problem. In some cases estrogen therapy was appropriate for those at highest risk. Nevertheless, osteoporosis was considered to be preventable if exercise and a diet with adequate calcium were followed. Yet much of the testimony about osteoporosis reflected weaknesses in that knowledge base. Even the statements about the preventive role of calcium supplementation did not rest on a firm evidentiary foundation. Perhaps the only dissent to the prevailing consensus was expressed by feminist activists and many female obstetrician-gynecologists who were distrustful of pharmaceutical companies and the emphasis on estrogen therapy as a remedy for osteoporosis, and of the interpretation of menopause as a deficiency disease.

RISK FACTORS

In light of the subcommittee's conclusions, the obvious goals for the future were to develop preventive strategies and efficacious treatments. The achievement of these goals required a detailed understanding of both those persons at risk for developing osteoporosis and those with the disease. The prevailing optimism notwithstanding, limited knowledge about the pathogenesis of osteoporosis and the cellular regulation of bone metabolism posed serious barriers to the development of preventive interventions. Nor was the therapeutic enthusiasm of this period based on carefully designed studies to evaluate efficacy. "There are few disorders in which it is more difficult to carry out appropriately controlled prospective trials for the evaluation of therapy," observed physician Lawrence G. Raisz and a colleague in 1985. Those postmenopausal women who suffered vertebral crush fractures (which generally occurred with minimal or no trauma) had to be followed for many years to determine whether the incidence of fractures was decreased by a specific regimen. It was even more difficult to determine therapeutic efficacy among those with hip fractures, if only because the incidence of a second hip fracture was not high enough to form the basis of a trial on patients with an initial fracture. While all women who lived long enough underwent menopause, only 25 percent of postmenopausal women developed vertebral crush fractures. A research agenda required knowledge about "the relative importance of low bone mass prior to the menopause, unusually rapid bone loss before or after the menopause, specific structural abnormalities, and changes in local and systemic regulators of bone growth and turnover."[12]

The interest in identifying risk factors en route to developing preventive interventions was by no means unique among osteoporosis researchers. A perennial theme in American history has been that disease may be prevented by following appropriate behaviors. Before the twentieth century, the advice offered by physicians grew out of their personal religious and moral beliefs. World War II brought a dramatic transformation. The prevention of disease and the maintenance of health no longer depended on personal beliefs; in their stead, medical science and epidemiology would provide guidance. More and more, the genesis of disease was attributed to the operation of risk factors. The public, in turn, was besieged by recommendations to modify behaviors that promoted disease. The assumption was that appropriate lifestyles held out the hope of preventing many diseases.

A series of highly visible and influential epidemiological studies seemed to provide evidence that individual lifestyles played major etiological roles. Ancel Keys's Seven Countries Study purported to show that the three critical variables in the genesis of coronary heart disease were age, blood pressure, and serum cholesterol. Dietary fats raised serum cholesterol and led to atherosclerosis. The Framingham Heart Study, launched in 1949, concluded that hypertension, obesity, cigarette smoking, and a family history of heart disease played crucial roles in coronary heart disease. In equally important studies, Austin Bradford Hill and Richard Doll demonstrated that smoking was the determinative element in the etiology of lung cancer. Doll and Richard Peto extended the argument when they claimed that two-thirds (and perhaps more) of all cancers were due to diet and smoking.[13]

In keeping with this trend, researchers attempted to explain the etiology of osteoporosis by identifying a series of risk factors and behaviors. Such knowledge would presumably then lead to the identification of behavioral and other interventions that might serve as the foundation for preventive strategies. During the early and mid-1980s, the search for risk factors began to accelerate.

Yet many of these studies revealed sharp disagreements within the osteoporosis community. It was equally clear that formidable obstacles remained in the design of studies that could identify pathological mechanisms involved in bone physiology. A majority of the claims about risk factors, therefore, relied on opinion or the experiences of individual clinicians. Equally important, the methodology employed to isolate risk factors involved correlations between variables. Such correlations could not demonstrate causality, however; at best, they could serve as a basis for future research. Nevertheless, associations of factors were often equated with causality, which in turn led to the belief that specific behavioral changes, in conjunction with medications, could mitigate risk.

In reviewing retrospective studies in 1984, for example, Lawrence G. Raisz and his colleague found that genetic or constitutional factors within one's body might be related to the degree of bone loss in early life. There was also a high concordance of bone mass in identical twins, and bone mass was consistently larger in African Americans than in whites. There was an association between low body weight and clinical osteoporosis, perhaps because the conversion of androgen to estrogen in body fat protected obese women, as well as the fact that increased weight could produce increased skeletal stress and thus preserve bone mass. Low calcium intake was another predisposing factor. Early menopause played an etiological role, as did the relative amount of exercise and weight-bearing activities. Raisz and his colleague also suggested that there might be an association between cigarette smoking and osteoporosis, which in turn was related to early menopause, while coffee drinking was associated with low bone mass.[14]

In their summary of risk factors a year later, a group of clinical epidemiologists led by Steven R. Cummings came to somewhat different conclusions. They agreed that early menopause was probably a more important determinant of bone mass than chronologic age, that sex and race played a role, and that thin women were more prone to sustain hip fractures than obese women. But they also found that geographical differences in hip fractures were striking. Data compiled from the records of the Mayo Clinic indicated that the United States had the highest rate (101.6 per 100,000 for women, and 50.6 for men), double that of Holland and Finland and more than six times that of Singapore. The reasons for such geographical variations were unknown. The role of heredity or reproductive history in regard to bone mass and susceptibility to fracture remained obscure. Although Western diets in general probably contained inadequate calcium, international and cross-cultural comparisons provided little support for the theory that calcium intake was an important determinant of cortical bone mass or the rates of hip fractures. The role of vitamin D in the pathogenesis of osteoporotic fractures was equally uncertain; the same was true of fluoride. Physical and weight-bearing exercises had positive effects, but the efficacy of specific types of physical activity had not been studied. Alcoholism and smoking clearly increased the risk of fractures. Chronic use of several types of medications (corticosteroids, thiazide diuretics) and a variety of medical disorders (diabetes, hyperthyroidism, hyperparathyroidism) accelerated bone loss. Falling was an especially important risk factor for osteoporotic fractures. The epidemiology of falling had not been well studied, although most falls and fall-related injuries occurred in and around the home.[15]

That same year Cummings reviewed fifteen case-control studies of patients who sustained hip fractures. Was there evidence that those who suffered hip fractures were significantly more osteoporotic than persons of a similar age who did not? The studies that protected against ascertainment bias (sampling bias) found smaller differences than studies that did not. Patients with hip fractures, Cummings concluded, "did not appear to be distinctly more osteoporotic than persons of a similar age." Factors besides bone mass, such as a tendency to fall, were possibly more important determinants of hip fractures. Bone mass measurements, then, might not be a reliable way to identify those at greatest risk of this type of fracture.[16]

Falling, of course, was a common event for all age groups. Among older persons—unlike their younger counterparts—falls

were from a standing height or less. According to Riggs and
L. Joseph Melton III (a colleague at the Mayo Clinic), the majority of age-related fractures were due to slips, trips, and "drop attacks" (caused by disturbances in blood flow to postural centers in
the brain stem), while a substantial minority resulted from sliding
off a bed or missing a chair while sitting down. The causes of
these falls were complex. Half were associated with organic dysfunction. Many aged persons had a variety of deficits, including
"diminished postural control, gait changes, muscular weakness,
decreased reflexes, poor vision, postural hypotension, vestibular
problems, confusion, or dementia." Certain specific diseases pre-
disposed some to fall. Iatrogenic problems (inadvertently caused
by a medical practitioner, treatment, or procedure), such as an excessive use of sedatives or overtreatment of hypertension, played a
role. Most falls, even among the aged, did not result in fractures.
What factors, then, enhanced fracture risks? Both bone loss and
trauma were necessary for fractures, but these factors were rarely
a sufficient cause of age-related fractures. The answer, Melton and
Riggs suggested, could be "an age-related reduction in protective,
energy-absorbing responses during a fall [such as an outstretched
hand] with a resultant increase in the actual trauma experienced."
They did not believe that it was possible to prevent falls; the solution lay in the prevention or delay of osteoporosis.[17]

Microstudies of risk factors gave rise to equally problematic
and controversial findings. In a comparison of fifty-eight women
suffering from postmenopausal osteoporosis (a crush fracture of
the spine) with an equal number of normal women, a group led by
John F. Aloia and funded by federal grants found that the former
had reduced bone mass in the skeleton and distal radius. These
researchers attributed the differences to smoking, early menopause, and inadequate calcium and vitamin D intake, and they
recommended that women classified as having a high risk of osteoporosis be given the choice of receiving estrogen-progesterone
therapy.[18] Another study of risk factors for fractures in the neck
of the femur in the elderly, one that involved a control group,
found that the population sustaining a fracture had less forearm

trabecular bone density, lower body weight, increased body sway, worse eyesight, and reduced mental acuity. But there was no evidence to indicate that vitamin D or estrogen deficiency played a significant role, or to suggest causality due to alterations in endocrine functions with respect to calcitonin or parathyroid hormone levels.[19]

There were daunting barriers in studying dietary factors for age-related bone loss and fractures. For one thing, changes in bone mass were relatively small and very slow. Moreover, bone mass was shaped by many variables other than diet, including one's age, sex, race, genetic constitution, muscular activity, and hormonal state. Medication use and smoking possibly played a role. Correlations between cortical bone mass and calcium intake were also weak. Very high and very low intakes of many nutrients might be harmful to bone, concluded A. Michael Parfitt in an article in *Lancet* in 1983, but variations in between those levels had "little effect." Calcium was the only nutrient that, when deficient, could be related to fracture risk. It was not the most important factor, although it was the easiest to correct.[20]

Two years later, an unsigned article in *Lancet* dealing with risk factors in postmenopausal osteoporosis came to somewhat different conclusions. It claimed that estrogen deficiency, which was associated with early menopause, was a "major pathogenic" factor, and that estrogen therapy protected against osteoporotic fractures. Moreover, dietary calcium might be a determinant of peak bone mass, while an impaired vitamin D metabolism had been found in those with osteoporosis. Physical exercise could conceivably be a preventive factor, while heavy alcohol intake was associated with an increased risk of fractures.[21]

Two epidemiological studies of hip fracture incidence during the early 1980s contradicted many of the studies that identified osteoporosis as being among the risks for this type of fracture. Both investigations were prepared by researchers at the National Institute on Aging, and both analyzed data from the National Hospital Discharge Survey (NHDS) from 1974 to 1979. The first study found that the age-specific incidence rate (per 100,000

white American females per year) rose from a baseline level of
about three fractures at ages 35–39 to twenty-two fractures at ages
40–44, a seven-fold increase. Thereafter the rate rose smoothly
with age, doubling every five to six years. There was no change
following menopause, which occurred between the ages 48 and
51. The authors cited a comparable study by B. Lawrence Riggs
and his associates, who—using DPA data—found that the regression line for diminishing bone density in normal women and
men from age 20 to 90 was linear. The study concluded by questioning the role of menopause as a key event in the pathogenesis
of osteoporosis, and urged that research efforts should focus on
the time prior to menopause.[22]

The second study, using the same NHDS data plus hospital
discharge data from the District of Columbia, found that the incidence rate for hip fractures among white women was approximately double that among white men, black women, and black
men at every age. The unusual susceptibility of white women,
therefore, was not attributable to either sex or race. Moreover,
international data seemed to confirm the observation that sex and
gender were not primary factors. The incidence of hip fractures
was the same among South African Bantu and Hong Kong Chi-
nese men and women. In Singapore, men with hip fractures out-
numbered women. In Jerusalem, the ratio of age-adjusted incidences of hip fractures in women compared with that of men was
highest for European Jews, lowest for Oriental Jews, and intermediate for the Israeli-born. Hormonal factors, therefore, could not
explain all of the variations in risk across racial and ethnic groups;
environmental and/or lifestyle factors, in addition to genetic factors, could possibly play a significant role in determining who
sustained a hip fracture.[23]

Despite differences among investigators, risk factors assumed
an increasingly important role in the debate over the pathogenesis
of osteoporosis. In a book addressed to a lay audience, David F.
Fardon provided a self-assessment list that individuals could use
to ascertain their risk of developing osteoporosis. Drawing on a
variety of medical sources, he listed twenty-five high-risk factors

and an equal number of low-risk ones. In addition to more common risk factors, he included fair skin; flat feet; scoliosis; and stomach, bowel, and periodontal disease.[24]

SCREENING

The search for risk factors in explaining fracture (especially hip fracture) rates took on new urgency with the development of more sophisticated technologies capable of quantifying BMD. By the 1980s there was increasing interest in screening for osteoporosis. The concept of screening for disease, of course, was by no means novel. In the late nineteenth century, inspections of school children centered around a search for contagious diseases that had a high mortality rate among this youthful population. By the beginning of the twentieth century, these examinations were expanded to include a search for general physical impairments in vision and other areas that interfered with the students' studies and required medical attention. During this same period, organizations such as the National Association for the Study and Prevention of Tuberculosis attempted to persuade people to check for early signs of tuberculosis and to have a chest x-ray, if warranted. Employers and insurance companies also advocated periodic physical examinations to find previously undetected diseases and to identify correctable harmful habits and poor hygiene. The movement to screen was given impetus by the physical examination of young males for military service during World War I, which found that more than half a million men suffered from physical and mental defects, many of which were preventable or remediable. The experiences with the draft during World War II were similar.

Screening, however attractive in theory, often had illusory results in practice, as illustrated by a famous effort in the early 1930s to use screening to identify which children should undergo tonsillectomies to reduce respiratory infections. A sample of one thousand New York City school children found that 61 percent had undergone tonsillectomies. The remaining 39 percent were then examined by a group of school physicians, who selected 45 percent from that group of children as candidates for surgery. Those youngsters still left over were examined by a second group of physicians, who recommended 46 percent of them for surgery. The contingent of unselected children was examined a third time, and tonsillectomies were recommended for 44 percent of that group. After three examinations, only sixty-five of the thousand children remained; they were not screened further because the supply of physicians not previously involved in the examinations had been exhausted.[25] Such variability was a reflection of the individualistic character of medical practice, the difficulties in identifying screening standards, and the problems in determining what constituted sufficient evidence to validate a particular therapy.

After 1945, Americans of all ages were urged to have annual health examinations. Public health advocates supported "multiphasic screening" to identify incipient chronic degenerative diseases, although issues regarding the effectiveness and costs of such procedures remained. Subsequently, individual screening tests for cervical, breast, and prostate cancer and other conditions, and for supposed markers of health or disease, such as cholesterol levels, became popular, even while their efficacy sometimes remained a subject of controversy.[26] Screening was linked with the concept of prevention, which was increasingly promoted during the last quarter of the twentieth century. Screening would reveal a disease in its earliest stage. The identification of required behavioral changes and the development of effective therapeutic interventions would follow, thus eliminating that disease as a threat. In one sense, faith in prevention was another reflection of the age-old dream of creating a disease-free society.

Under these circumstances, it was understandable that screening for osteoporosis would arouse the interest of clinicians. Screening would presumably identify those persons at high risk for fractures, who would then undergo preventative treatment. Yet advocates of screening faced considerable obstacles. The efficacy of screening generally depended on several variables. The first was to identify the at-risk population. The second was to have accurate, reliable, safe, and cost-effective means of screening.

Finally, therapies of proven efficacy had to be available; otherwise screening would have little or no value. In the 1980s the latter requirement had not been met. The evidence for the effectiveness of existing therapies for osteoporosis, such as estrogen-progesterone and calcium supplementations, was by no means conclusive. Nor had the measurement of BMD by the various screening technologies led to the adoption of a clear numerical scale to identify osteoporosis.

Equally important, the very definition and diagnosis of osteoporosis was still contested. B. E. C. Nordin, for example, pointed out that spontaneous vertebral compression, precisely because it was spontaneous, denoted the presence of severe osteoporosis. This was not true of other fractures, however, nearly all of which involved trauma. Osteoporosis increased the risk of fractures, Nordin argued, but it did not cause a fracture per se. Nordin emphasized that all women and men, if they lived long enough, would become osteoporotic. Age, therefore, was a critical variable in developing a risk-measurement scale. Blood pressure, for example, rose according to one's age, and hypertension affected everyone to some degree. This did not prevent physicians from defining normal blood pressure in terms of the range for young adults. But in assessing its significance for an individual, that person's age had to be taken into account. The same was true for osteoporosis. The clinical significance of the measurement was a function of the age of the subject. Thus a value that lay within the normal range for a subject's age meant something different from a value that was low for that person's age. Hence the latter measurement level, according to Nordin, might be termed "accelerated osteoporosis," and the former "simple osteoporosis." Nordin's analysis thus suggested that screening technologies presented complex methodological issues.[27]

Before 1980, screening for osteoporosis had been a relatively minor issue. In the mid-1980s, however, debates over the wisdom and desirability of screening for this condition became more common. Aside from the scientific merits of screening, other considerations played a role. In 1984 only about twenty-five clinics offered SPA screening; two years later the number rose to approximately five hundred.[28] Equally significant was the fact that companies producing screening machines were engaged in an extensive marketing campaign. One physician complained that the manufacturers' sales forces were "inundating internists, orthopedists, and radiologists with financial data demonstrating the lucrative nature of mass screening but without solid scientific data indicating that it should be undertaken."[29] The marketing of this relatively new technology and its accompanying publicity had the effect of heightening public and professional concern with what was presented as a major public health problem. While medical data played a role, screening was also driven in part by the sales strategies of companies that manufactured the equipment.

Was it desirable for women to have non-invasive measurements of bone mass as a routine screening procedure? The answer to this question, according to Cummings, depended on four criteria. First, did the condition cause sufficient mortality or morbidity to warrant routine screening? Second, was treatment during the asymptomatic phase effective in reducing morbidity or mortality? Third, was the screening test reasonably inexpensive, safe, and acceptable to patients? Finally, did the screening test provide an adequate predictive value?[30]

The answers to such questions were by no means simple. There was little doubt that the morbidity and mortality caused by hip fractures warranted screening, but the measurement of bone mass was "of uncertain value" in assessing the risk of such fractures. Wrist fractures caused only transient disability, and their risk could not be predicted. Serial measurements of bone mass to estimate the rate of bone loss were also imprecise and increased the costs of screening. Cummings concluded that his findings "do not support the routine use of non-invasive measurements of bone mass in perimenopausal women." Moreover, 80 percent of all hip fractures occurred after age 70, and factors other than a reduction in bone mass were important in determining who would sustain a fracture, especially in those who experienced "proneness to falling and the inability to protect themselves from injury during a fall."[31] In posing the same questions as Cummings, two Johns Hopkins School of Medicine physicians came to similar conclusions. To screen postmenopausal women would merely document what was already known, namely, that women rapidly lose bone mineral after menopause. For older women, any other form of age-related bone loss was relatively low. The use of SPA and DPA, therefore, was justified only in research protocols.[32]

Others echoed Cummings's analysis. In an editorial for the *Annals of Internal Medicine*, Susan Ott pointed to the absence of longitudinal studies (repeated observations of the same variables over long periods of time) on measures of bone loss in a cohort of women. How, then, could a 45-year-old women know her risk of developing fractures? The new bone-screening technology admittedly demonstrated great promise, but its value for predicting fractures remained in doubt. Longitudinal studies and methods of assessing the quality and the quantity of bone were necessary. Hence, Ott stated, screening was "premature, albeit profitable."[33]

Still others expressed tentative opinions, hedged with qualifications, about the wisdom of routine screening. In an article on involutional osteoporosis in the *New England Journal of Medicine*, Riggs and Melton were equivocal in their attitude. They expressed concern with the proliferation of screening centers that measured bone density on an indiscriminate "walk-in" basis. Such measurements were expensive, and screening millions of those at risk was not a defensible use of the nation's health care resources. A more cost-effective approach might be to use historical risk factors, although it was not yet possible to weight these factors according to their relative importance. Densitometry was less useful for women over age 75, because most of them were below the fracture threshold, and women with and those without fractures hardly differed in their bone densities.[34] The American College of Physicians' report on BMD emphasized the absence of prospective, controlled studies identifying densitometry as a basis for prescribing specific preventive measures against fracture risk. Hence this organization was opposed to general (mass) screening.[35]

The rapid proliferation of the screening technology for BMD

also raised financial concerns. The Health Care Financing Administration, which had decided not to permit reimbursement for bone mineral screening in patients receiving Medicare, was under considerable pressure to reverse its decision. Such pressure had already forced some third-party payers to reimburse charges for osteoporosis screenings. Yet studies showed a weak correlation between hip fractures and bone mineralization. "Widespread uncontrolled screening for osteoporosis," according to a sounding-board article in the *New England Journal of Medicine*, "would contribute little to our understanding of the disease or its treatment." It would not tell at which specific levels or at which interval changes treatment should be initiated, nor would it provide information about optimal estrogen treatments or other therapies. At best, screening was a useful research tool and could be helpful in specific clinical situations. Large-scale screenings, however, were not warranted. Responses to this article were generally favorable.[36]

Nuclear-medicine physicians who provided bone mineral densitometry, and their professional organizations, were quick to offer rebuttals. The Society for Nuclear Medicine and the American College of Nuclear Physicians insisted that DPA was effective, and that the failure of third-party payers to reimburse practitioners for this procedure would result in millions of hospital admissions and considerable pain.[37] Densitometric screening, wrote two San Diego radiologists in response to the Cummings and Ott articles, had merit even if its only consequence was to make the public aware of a problematic condition. They saw osteoporosis as a disease that had to be attacked in adolescence. Although screening for increased spinal or distal radius fractures might not be justified from an epidemiological standpoint, they argued, it certainly was from a personal viewpoint. Finally, the radiologists were distressed by the implication that densitometric screening was undertaken from a profit motive. "Medicine should never be a business," they insisted, "and it is our conviction that every woman has a right to know truthfully the risk of fracture at important [body] sites."[38]

In late 1988 the National Osteoporosis Foundation's Scientific Advisory Board submitted a long document to the Health Care Financing Administration, laying out its own guidelines for bone mass measurements. Osteoporosis and its associated fractures were not only an important public health problem, but also one that would worsen in the future as America's general population grew older. Accurate, precise, and safe methods were available to measure bone mass, which in turn could guide appropriate clinical decisions and lead to improved health. The board conceded that there were not enough data to justify the use of bone mass measurements to screen unselected people, to identify rapid bone losers, or to monitor therapy. Nevertheless, sufficient data existed to support the use of bone mass measurements in four situations. First, to diagnose significantly low bone mass in estrogen-deficient women, in order to make decisions about HRT. Second, to diagnose spinal osteoporosis in patients with vertebral abnormalities or with osteopenia that showed up on x-rays. Third, to monitor and adjust treatments in patients receiving long-term glucocorticoid therapy. Finally, to diagnose low bone mass in patients with asymptomatic hyperparathyroidism, in order to identify those at risk for severe bone loss who might be candidates for surgical intervention.[39]

NOF's Scientific Advisory Board insisted that BMD was highly correlated with bone strength, which in turn was a determinant of susceptibility to fractures. Bone mass could be measured accurately by SPA, DPA, and quantitative computed tomography. These measurements were not a diagnostic test for fractures; instead, they measured a risk factor (reduced bone mass) for future fractures and could properly be used for risk stratification. Board members conceded that there were other risk factors for fractures, especially those related to falling. Trauma was a minor cause in vertebral fractures, while falls played a significant role in hip fractures. Yet the risk of falls only increased from 19 percent per year among women at age 60–64 to 33 percent per year at age 80–84. Falling, therefore, could not explain this exponential increase in hip fractures, although a change in the type of falls might have an effect. Risk factors based on an individual's history or a physical examination were not reliable predictors of bone mass, nor could a person's fracture risk be accurately estimated by assessing risk factors. The advisory board, despite diluting the force of its statements with many qualifications, still placed a great deal of faith in bone mass measurement technology and ERT. "There is compelling evidence," its members reported, "that long-term estrogen therapy prevents bone loss and fractures."[40] In future years, as we shall see, the NOF played an increasingly visible role in publicizing the dangers of osteoporosis and the importance of screening, and in promulgating standards for clinical management of the disease.

The analysis and recommendations of the NOF's Scientific Advisory Board were not universally accepted. The board ignored some data that ran contrary to their recommendations, notably the role of falls in the etiology of fracture. Moreover, the use of bone density measurements without due consideration of a patient's complete medical status sometimes led to inappropriate interventions. In their review of 800 consecutive patients age 31–86 at an osteoporosis center at the University of Kansas Medical Center, a group of physicians noted that 83 out of 180 had a contributing diagnosis. In these 83 patients, no fewer than 128 medical problems other than osteoporosis were detected. Many of these patients had rheumatoid arthritis, chronic obstructive pulmonary disease, or asthma. Glucocorticoid therapy, which contributed to bone loss, was common for many of these conditions. Osteoporosis also appeared with increased frequency in patients with scoliosis and with certain gastrointestinal disorders. To ignore confounding diagnoses meant that bone density measurements were occurring in a medical vacuum.[41]

More importantly, the information obtained by bone mass measurements often added to, rather than resolved, controversies. Although routine bone mass screening was gaining favor toward the end of the 1980s, weaknesses in the knowledge base about bone physiology persisted. Bone loss, for example, was a complex phenomenon, and simple bone mass measurements did not shed much light on bone physiology. For example, in a longitudinal study of 139 normal women (age 20–88), Riggs and his colleagues found sharp differences in bone loss in the spine compared with the peripheral skeleton. In the former, there was considerable premenopausal bone loss, but no acceleration at menopause. In the latter, bone loss did not occur until after menopause. Hence menopause could not be the only cause of vertebral osteoporosis.[42]

Disagreements were common. Was bone loss a continuous process, or did it accelerate after menopause? One claim was that the best predictor of a high level of bone mass in later life was high bone loss in one's early adult life. This concept had not been validated, but, even if true, the idea of fast and slow bone losers was not relevant to the main argument. The issue of bone mass in the elderly was also a contested one. Should bone mass in an individual be compared with an age-matched control person? Such a comparison was problematic, if only because virtually all people over the age of 75 had bone mass values that fell into the at-risk range for fractures. Bone mass measurements, concluded Ignac Fogelman (a member of the Department of Nuclear Medicine at Guy's Hospital in London), "have created more problems than they have solved. There is a considerable amount of conflicting data in the literature." Nor had there been "real progress in the treatment of osteoporosis." Fogelman noted that what was needed were better-planned long-term clinical studies.[43] Identifying risk, wrote Robert Lindsay, remained "a major problem." Risk-factor assessment and bone mass measurement, used in combination, offered "the closest approximation to the ideal, although it is not yet clear how good identification is."[44]

A consensus statement on prophylaxis and treatment, prepared at an international conference (which included leading American researchers) sponsored by the European Foundation for Osteoporosis and held in Denmark in 1987, only added to the confusion. The public report endorsed estrogen treatment for postmenopausal women and touched on other regimens: calcium, exercise, fluoride, vitamin D, anabolic steroids, and calcitonin. It concluded by emphasizing the need for additional knowledge.

Research was required to illuminate the factors that determined the accumulation of bone tissue during the period of the body's growth and maturation, the loss of bone tissue thereafter, and the strength of the adult skeleton. To identify individuals at special risk for developing osteoporosis and fracture was equally important, as was the development of precise methods to monitor the effects of treatment. Finally, there was a need for preventive and treatment strategies that were safe, effective, inexpensive, and acceptable to patients.[45]

Yet, as Tony Smith (deputy editor of the *British Medical Journal*) subsequently pointed out, the conference did not deal with a vital question: which women should be treated? A far more hostile objection came from Nordin and other knowledgeable individuals who had been invited to the conference. The consensus statement, they charged, represented only the views of panel members; the invited experts were neither consulted nor shown the final document. Nordin and his fellow critics pointed to the ambiguity of the calcium recommendation and a mistaken emphasis on the side effects of anabolic steroids. The failure to identify which women should be given estrogen, moreover, constituted a major weakness in the conference report. In their view, the primary determinant of bone density over a period of many years after menopause was the initial BMD value at menopause. Thus the women who required preventive treatment were those whose bone density was low when menopause began. The dissenters were also critical of the emphasis that Tony Smith placed on such traditional risk factors as a small body build, menopause, and one's family history; they claimed that evidence to support the role of these factors was lacking.[46]

THERAPY

The debate about the importance of risk factors and screening was accompanied by equally contentious discussions about therapies that would, if not restore lost bone, at least prevent further losses. Although stated in authoritative terms, the therapeutic recommendations offered to the public by physicians—even those who enjoyed a prominent position within the osteoporosis research community—often lacked a solid evidentiary foundation and represented instead their individual experiences as clinicians. Therefore, laypersons had little to fall back on other than the recommendations of their personal physicians, who, in turn, were often influenced by representatives of the pharmaceutical industry and clinicians who enjoyed close relationships with these firms.

In the 1980s discussions about therapy generally revolved around the role of estrogen, given the fact that there were few other alternatives available then. The use of estrogen, except as a preventive therapy for osteoporosis, was hotly contested during this decade. An increasingly vocal feminist movement was critical of the medical model that defined menopause as a deficiency disease, gave physicians (often male) a leading role in dealing with women's health, and promoted the use of estrogen as a therapy that would preserve youth and vitality. Yet for many women estrogen was—to use Elizabeth S. Watkins's term—an elixir, and they remained largely unaffected by those who insisted that menopause was a perfectly normal phase of life. Such popular women's magazines as *Red Book* and *Ladies' Home Journal*, which, respectively, reached fourteen and twenty million readers each month, published enthusiastic articles about the anti-aging and osteoporosis-prevention qualities of estrogen. Nor was the publicity accorded to estrogen limited to women's magazines; *Newsweek* and other general publications followed suit.[47]

The use of estrogen by menopausal and postmenopausal women was clearly on the rise during the 1980s. In the preceding decade, fears that this drug elevated the risk of endometrial cancer had led to a decline in its use. By 1980 prescriptions for estrogen had fallen to 14 million. Between 1983 and 1992, however, sales rose from 17.8 to 31.7 million. The estrogen skin patch, which came on the market in 1986, had sales of 4.7 million five years later. The number of progestin prescriptions (intended to minimize the risk of cancer) enjoyed equal popularity: in 1992, 11.3 million prescriptions were written. Pharmaceutical firms, which spent large sums marketing and promoting such drugs, were the beneficiaries of such sales.[48]

Ayerst Laboratories was undoubtedly the most aggressive firm in marketing estrogen. Premarin, which it introduced in 1942, was the best-selling drug of all time. In the 1960s ads for this drug emphasized that it would enable older women "to age physiologically rather than pathologically." Estrogen would help prevent a loss of skin elasticity, osteoporosis, and even depression, to cite only a few of its touted benefits. A decade later publicity for this drug emphasized the scientific legitimacy and clinical necessity of estrogen. A multipage advertisement in *JAMA* in 1977 included a photograph of a gray-haired, dejected woman. The caption was revealing: "*THE MENOPAUSE*. Does she just have to live with it?" Ayerst also employed a large sales force to ensure that clinicians would prescribe Premarin above all other estrogen products. While this company was not alone in paying speakers from the medical profession to tout the superiority of their estrogen formulations, no other pharmaceutical firm could match the extraordinary success of Ayerst in persuading both women and their physicians about the superiority and effectiveness of Premarin.[49]

Generally speaking, the use of ERT to prevent bone loss drew broad support among clinicians. At a medical-staff conference held at the University of California in 1985, Richard Haber (chief of the university's Division of General Internal Medicine) expressed his belief that estrogen deficiency played a causal role in the development of postmenopausal osteoporosis and fractures, and that estrogen therapy could prevent these events. For Haber, the benefits clearly outweighed the risks, which he thought were minimal. For women unwilling or unable to use estrogens, there were "no demonstrated therapeutic strategies that have equivalent bone-preserving properties." Supplemental calcium therapy was an alternative, but studies did not always show it to be better than a placebo.[50]

Most articles in the latter half of the 1980s echoed Haber's views. To be sure, the mechanism for estrogen's action on the skeleton was undetermined, largely because investigators could not find estrogen receptors in bone. The net effect, however, was improvement in the efficiency with which the body handled calcium homeostasis. One popular theory was that estrogens stimulated the production of calcitonin, which then was responsible for a reduction in bone resorption. The evidence for this theory, however, was either circumstantial or inconclusive. Nonetheless, the available data, noted Lindsay, suggested that the "judicious use of estrogen therapy can reduce fracture incidence in aging women."[51] Few disagreed with the advice offered by a group of Mayo Clinic physicians that ERT "should be instituted as soon after menopause as possible and seems to be well tolerated until at least 75 years of age."[52] Physicians were told that they had an obligation to thoroughly explain the benefits and risks of HRT, thus permitting patients to participate in the decision-making process. Such participation, of course, would "prevent fear and ignorance from depriving many women of a treatment that can significantly improve the quality of their later years."[53]

Estrogen use also seemed to have a beneficial impact on the all-cause mortality rate. One study followed 1,016 women— placed on estrogen following a hysterectomy—for 14,318 patient years. This investigation found that there was a marked drop in deaths from all causes over those that otherwise might have been expected. There had also been a significant decrease in the clinical evidence of osteoporosis in these women. While they experienced an increase in breast cancers over those that might have been expected, fewer of these women died from breast cancer than the anticipated mortality rate.[54] A Lipid Research Clinics Program's prevalence study of cardiovascular disease among a like group of women came to a similar conclusion. The mortality rate among estrogen users, irrespective of their hysterectomy status, was significantly lower than that among non-users (3.4 vs. 9.3 per 1,000). The inverse association of estrogen use and a risk of death suggested that estrogen increased the level of high-density lipoprotein (HDL) cholesterol, which in turn was associated with a reduced risk of coronary heart disease (the major cause of death in postmenopausal women). The investigators, however, conceded that observational studies (such as the one they conducted) had limitations, and they called for more controlled experimental investigations.[55]

In an AARP book published in 1988, Peck and Louis V. Avioli devoted a chapter to estrogen therapy. A postmenopausal woman, they wrote, would spend more than a third of her life with lowered levels of estrogen. The reduction in estrogen led to three changes in a woman's health: varying degrees of disturbing symptoms; bone tissue loss; and perhaps susceptibility to heart attacks, caused by a rise in low-density lipoprotein (LDL) cholesterol and a decline in HDL cholesterol. Estrogen therapy was the most effective way to relieve disturbing symptoms and prevent osteoporosis. It was clear that there were circumstances under which estrogen therapy could not be used, including in women who had breast and endometrial cancers, blood clots, uterine fibroid tumors, and several other rare conditions. Nevertheless, the benefits were obvious. Numerous studies demonstrated that estrogen-treated postmenopausal women were far less likely to break a bone than untreated women. A high calcium intake and exercise were not effective substitutes. Estrogen therapy, Peck and Avioli insisted, "is an investment in bone health—it has long-lasting benefits."[56]

Riggs was much more restrained in his therapeutic recommendations. Too many physicians and patients failed to understand that pain, especially from new vertebral fractures, subsided in weeks or a few months, even in the absence of drug therapy, he noted. Chronic pain due to spinal deformity could persist even if drug therapy favorably altered the balance between resorption and bone formation. In both cases physical therapy, heat, massage, analgesics, and occasional bed rest sufficed. Insofar as the drugs used to treat involutional osteoporosis—estrogen, calcitonin, calcium, and vitamin D—the best result that could be expected was to maintain bone mass or to slow the rate of bone loss. Older persons with senile osteoporosis had already lost most of the bone they would lose, and there was little evidence that they would benefit from estrogen or calcitonin therapy. Moreover, the age of these individuals and their increased incidence of asymptomatic atherosclerosis increased the dangers to them from estrogen therapy. "All available regimens for the treatment of osteoporosis," Riggs wrote, "are only partially successful and are ineffective in some patients." Prevention was "the only cost-effective approach." Nevertheless, there was no agreement on how high-risk individuals could be identified. The use of historical risk factors was inexact, albeit useful. Bone density measurements were expensive, and it was not cost-effective to screen the millions of people who were potentially at risk. Riggs's sober discussion suggested that much of the therapeutic enthusiasm of the 1980s was somewhat premature.[57]

The debate about the efficacy of estrogen was by no means unique. Disagreements over other therapies were also common. There were differences regarding the role of calcium and calcium supplementation. Avioli, for example, believed that the amount of calcium consumed in childhood and adolescence was a major determinant of one's bone mass content. He also believed calcium supplements increased the efficiency of estrogen therapy and physical exercise and retarded the rate of cortical bone loss in postmenopausal women. Too many women had calcium intakes well below the recommended daily allowance. Avioli therefore supported the conclusion of a 1984 NIH Consensus Conference, which stated that increasing one's calcium intake by 1,000–1,500 mg per day would reduce the incidence of osteoporosis in postmenopausal women.[58]

The year after Avioli published his piece, J. Chris Gallagher (an associate of Heaney at Creighton University's School of Medicine) expressed a completely different view. "There are virtually no studies of the effect of calcium on the bone density of osteoporotics." Gallagher noted that at the beginning of treatment, calcium supplements probably inhibited bone resorption, but they did not affect bone formation. Moreover, this effect only lasted for several months, and it was followed by a reequilibration between resorption and bone formation. It was equally unclear what calcium-requirement figure was appropriate for osteoporotics.[59]

Such differences of opinion were by no means uncommon. The treatment of osteoporosis, wrote two physicians at New York's Hospital for Special Surgery, "is both complex and controversial." Much work, they added, "remains to be done . . . and many questions remain."[60] The use of calcitonin, fluoride, vitamin D, anabolic steroids, and parathyroid hormone were equally controversial therapies, in part because of the paucity of long-term clinical trials. Calcitonin was known to inhibit bone resorption by preventing calcium release from bone. Its level decreased with age, particularly among women. Calcitonin supplementation thus seemed to have an obvious therapeutic rationale for its use in osteoporosis. Nevertheless, the initial increase in bone mass during the first two years of this treatment was followed by bone loss during the third year of calcitonin therapy. Thus this therapy was perhaps of limited usefulness, because its ability to inhibit bone resorption was lost due to the "escape phenomenon"—the densensitization of osteoclasts to calcitonin. This was by no means an unfamiliar phenomenon; bacteria also became resistant when antibiotic therapy was overused.[61]

Some therapies underwent cycles of enthusiasm and disillusionment. The use of sodium fluoride to treat osteoporotic individuals was stimulated by findings that potable water containing fluoride in doses of four to five parts per million resulted in an increased resistance to dental caries and fewer vertebral fractures. It was well known that the human skeleton accumulates fluoride with age. Hence there was considerable optimism that fluoride therapy could be used to prevent vertebral crush fractures. In the early 1960s, Clayton Rich and his colleagues began to treat osteoporotic patients with sodium fluoride in the belief that it would strengthen the skeleton but not lead to other changes. They found that this treatment increased calcium retention.[62] Subsequent research, however, revealed that fluoride therapy also resulted in increased osteocyte resorption and frequent microfractures in old bone.[63] Consequently, interest in this therapy declined. In the 1970s and 1980s there was renewed enthusiasm and support for fluoride therapy, as studies seemed to demonstrate that the dosage level and the simultaneous administration of calcium supplements played a critical role. Nevertheless, the side effects of this treatment were significant, including anemia, gastrointestinal problems, arthritis, and an increased incidence of fractures. The FDA failed to authorize its clinical use, despite the fact that it had been approved by regulatory agencies in eight European countries. An FDA advisory committee convened in 1989 concluded that fluoride had yet to demonstrate its efficacy in the treatment of osteoporosis.[64] Shortly thereafter, two prospective, randomized, double-blind, placebo-controlled clinical trials came to the conclusion that sodium fluoride therapy with calcium supplementation was no more effective than a placebo in treating postmenopausal osteoporosis. "There is no role for NaF [sodium fluoride] in the treatment of osteoporosis outside the confines of clinical research," concluded the authors of one of the studies.[65]

While faith in estrogen therapy was peaking in the 1980s, investigators began to pursue other lines of research. There was growing interest in bisphosphonates (once known as diphosphonates). Synthesized more than a century earlier, bisphosphonates, like many other chemicals, were first developed for industrial uses. They were utilized as antiscaling and anticorrosive agents in a variety of products, to prevent the deposition of calcium carbonate scale. In the 1960s, they entered the medical armamentarium as experimental drugs. The therapeutic rationale for their use in treating metabolic bone diseases was that bisphosphonates inhibited bone remodeling, particularly bone resorption, by a variety of mechanisms. In the case of Paget's disease, bisphosphonates had a demonstrated therapeutic efficacy. There were, however, few studies of their effectiveness in treating osteoporosis. In 1976 Heaney and his colleague Paul D. Saville observed a decrease in both bone resorption and bone accretion in a group of postmenopausal osteoporotic women treated with etidronate disodium (EHDP), a drug in the bisphosphonate class. The decrease in mineralization (accretion) proved to be disappointing, and the mechanism of its action remained uncertain. Heaney and Saville expressed the hope that a lower dose of EHDP might reduce bone accretion less than it reduced bone resorption, and they urged further experimentation with other treatment regimens. In their eyes EHDP was "both a potent and a well-tolerated bone-active drug."[66]

More than a decade later, bisphosphonates remained as experimental therapeutic agents for osteoporosis, but they were not yet approved as a drug treatment for osteoporosis (although EHDP had received approval for the treatment of Paget's disease). The precise dosage and most effective method of administration (continuous or intermittent) were unknown. Bisphosphonates, according to one authority, had the potential to be effective as an antibone resorber and thus prevent osteoporosis.[67] The preoccupation with estrogen therapy, however, deflected interest from pursuing other alternatives. Within a few years, a variety of reasons for concern about the safety of estrogen therapy would begin to undermine its popularity. At the same time, one of the bisphosphonates would receive FDA approval and become the drug of choice in the treatment of osteopenia and osteoporosis.

By the close of the 1980s, osteoporosis had assumed an important place in the pantheon of diseases that posed major threats to the American population. The NOF and the AARP were working together assiduously to increase public awareness of osteoporosis. In 1988 the AARP published Peck and Avioli's *Osteoporosis: The Silent Thief.* Their rationale was clear: the growing numbers of persons over the age of 65 would reach 20 percent of the American population by 2030. This development meant that age-associated disorders—arthritis; vascular diseases of the heart, brain, and limbs; cancer; and Alzheimer's disease—would increase. Unlike many of these chronic diseases, osteoporosis was potentially preventable. Peck and Avioli stated that their book was designed to bring the findings of researchers to the public's attention, including the causes of osteoporosis, indications of which persons were at risk, and ways in which the disease could be prevented. Peck and Avioli pointed to new ways of measuring bone tissue that shortly would be available to all, and they attempted to clarify "myths" that inhibited therapy and prevention (e.g., the myth that estrogen caused cancer). The book's fourteen chapters provided a comprehensive overview of osteoporosis, and it was designed to appeal to a broad public. Its authors—like others associated with foundations devoted to particular diseases—looked forward to a bright future. "There is every reason to hope," Peck and Avioli wrote, "that osteoporosis can be eliminated within the next twenty-five years, with proper attention to research."[68]

The desire to educate the public was accompanied by an effort to bring research findings to physicians and other health care providers. In all likelihood, the majority of physicians in such specialties as internal and family medicine, obstetrics and gynecology, and geriatrics were not familiar with the journals in which most of the articles on osteoporosis were published, even though doctors in these fields were the largest set of medical-service providers for osteoporotic patients. To reach this particular audience, Avioli persuaded leading figures in the osteoporosis research community to author chapters in a volume summarizing the knowledge base at that time. Contributors to the first edition of *The Osteoporotic Syndrome: Detection, Prevention, and Treatment*, which appeared in 1983, included Nordin, Melton, and Heaney. Within a decade, two more editions were published.[69]

Avioli's book was by no means unique. In 1988 Riggs and Melton edited a much larger and far more detailed volume that ran to 500 pages and included chapters by virtually every leading figure in the field. Their book, entitled *Osteoporosis: Etiology, Diagnosis, and Management*, was divided into four sections: basic science, clinical aspects, pathophysiology, and management. Seven years later a second edition appeared. Both editions were directed to general physicians, specialists, residents, and researchers. "Despite belated understanding of the enormity of the problem of osteoporosis," Riggs and Melton wrote in the preface to the second edition, "prospects have never been better for bringing the disease under control."[70]

Notwithstanding the optimism expressed by Riggs and Melton, formidable problems in understanding bone physiology remained despite nearly a half century of research. "Remarkably little is known," wrote Roberto Pacifici and Avioli in 1993, "regarding the underlying pathological process(es) that undermine matrix and cellular aspects of skeletal tissue in the aging, fracture-prone female." Was this age-related loss of bone a consequence of senescence that simply reflected an age-dependent dysfunction of bone cellular activity? Did a "discordant cybernetic 'couple' between the osteoblast-osteoclast cellular components and an accumulation of mast cells" stimulate bone resorption? Given differing incidence and prevalence rates of hip fractures throughout the world, what were the roles of "genetic predisposition and race, latitude and degree of sunlight exposure, physical activity, excessive alcohol, caffeine, nicotine, low calcium intake, and nutritionally inadequate diets"?[71] Was osteoporosis a pathological diagnostic category or simply the result of aging? That such questions were raised, to say nothing about debates over the efficacy of a variety of therapeutic interventions, suggested that knowledge about the etiology, pathology, and treatment of osteoporosis left much to be desired.

In the keynote address delivered at a 1972 conference on metabolic bone disease that was attended by 500 scientists, clinical investigators, and clinicians from twenty-nine countries, Charles E. Dent (of London's University College Hospital Medical School) expressed the belief that studies of osteoporosis continued to remain in their "usual muddled state." The scientific community was "obsessed" with the belief that osteoporosis was caused by defective calcification mechanisms. "Old age, with its inevitable slow insidious osteoporosis," he noted, "seems to me to be different, perhaps not a real disease." Dent believed the focus should be on ways to build the best possible skeleton during childhood, when stresses and strains clearly produced a larger, stronger skeleton. "How far it can be said that 'senile osteoporosis is a paediatric disease' and how far activity in later life would also be beneficial clearly needs further study, presumably by epidemiological methods, for slow changes in rates of bone formation and dissolution, sufficient to cause big effects in 40 years, are quite unmeasurable now and in the foreseeable future."[72]

Doubts about weaknesses in the knowledge base about osteoporosis, however, were generally confined within a relatively small specialty group. Outwardly, its members insisted that osteoporosis was a serious public health problem. Women were urged to consult their physicians about using appropriate therapies and modifying their lifestyles to minimize risk. Their physicians, particularly those in clinical practice, in turn relied on both pharmaceutical company representatives and members of the osteoporosis research community for information to convey to their patients; most lacked the time (or the knowledge) to evaluate a body of literature that was growing ever larger year by year. Doubts and disagreements rarely entered into this clinical relationship. The result was a confirmation of the reality and legitimacy of osteoporosis as a diagnostic category, and its elevation as a major health hazard.

CHAPTER FIVE

Internationalizing Osteoporosis

"Despite the current prominence of AIDS," wrote Robert P. Heaney in 1991, "osteoporosis may well be the disease of the 1990s." Osteoporosis, along with Alzheimer's disease, had the potential to become a budget buster. Even if disease prevalence among the aged remained constant, Heaney noted, the rising numbers in this segment of the U.S. population would increase medical costs many fold in future decades. More alarming were European data indicating a sharp rise in the age-adjusted incidence of hip fractures during the previous thirty or forty years.[1] Heaney's warnings were echoed by L. Joseph Melton III (of the Mayo Clinic). He predicted that the upsurge in the number of the elderly worldwide would result in 6.26 million hip fractures by 2050. This exponential rise was due to an age-related increase in the risk of falling and an age-related reduction in bone strength.[2]

Such views were indicative of the growing belief that osteoporosis was as much an international problem as a domestic one. In its origins in the mid-twentieth century, the diagnosis, at least in its modern form, first achieved prominence in the United States and, to a lesser extent, in Great Britain and Australia. This is not to imply that osteoporosis was of little or no interest elsewhere. It is simply to suggest that a preoccupation with bone research was concentrated in the United States, which emerged after World War II as the preeminent nation in medical and scientific research.

By the 1980s, however, a dramatic transformation was taking place. A mutually recognized community of osteoporosis researchers and clinicians had come into existence, and their work quickly transcended national borders. In succeeding decades osteoporosis would be redefined as a problem faced by all nations with older populations. Newly founded organizations in Europe and the World Health Organization (WHO) began to promote medical and public awareness of this disease. In the United States, the NOF expanded its activities by not only stimulating public awareness, but also publishing a manual for clinicians that laid down guidelines for the diagnosis and treatment of osteoporosis. Federal health agencies added to the chorus of voices elevating the importance of osteoporosis. During these same years pharmaceutical companies not only developed and aggressively marketed a variety of new drugs, but they also sponsored epidemiological surveys that purportedly demonstrated the seriousness and pervasiveness of osteoporosis among older women and the risk factors that played a role in this condition. Their financial support of organizations and clinicians served to strengthen the growing influence of these firms and expand the market for their drugs.

ROLE OF ORGANIZATIONS

In the 1980s an organizational response to problems relating to osteoporosis began to take shape, one that transcended national borders. Several NIH Consensus Development Conferences had attempted to lay down therapeutic guidelines. Three years after the founding of the NOF in 1984, the European Foundation for Osteoporosis (EFFO) was formed. It was followed by the International Federation of Societies on Skeletal Diseases in 1995, with a membership consisting of seventy-four national osteoporosis societies from forty-six countries. Recognizing the need for a single global organization to bring together scientists, physicians, corporate partners, and patient advocacy groups, in 1998 the EFFO and the International Federation of Societies on Skeletal Diseases agreed to consolidate. At the European Congress on Osteoporosis in Berlin that year, the merger was completed with the founding of the International Osteoporosis Foundation (IOF). In the mid-1980s, periodic international osteoporosis symposiums began to be held. The Health Council on Osteoporosis was formed in 1991 in recognition of the need for worldwide educational programs and the dissemination of advances in the diagnosis, prevention, and treatment of this condition.

Before 1990, researchers published their findings in a variety of medical and scientific journals, including *Calcified Tissue International*, the *Journal of Bone and Mineral Research*, the *Journal of Clinical Endocrinology and Metabolism*, and more general medical publications, such as *Lancet* and *JAMA*. Clinicians who dealt with osteoporosis came from diverse backgrounds: family practice, gynecology, rheumatology, orthopedics, geriatrics, radiology, and endocrinology. Articles dealing with clinical advances in osteoporosis appeared only sporadically in journals within those specialty areas. Many clinicians, however, did not subscribe to journals outside of their specialty. To ensure a more unified approach, the Osteoporosis Society of Great Britain, the EFFO, and the NOF collaborated to create a new journal. The first issue of *Osteoporosis International* appeared in 1990, and it quickly became the major outlet for articles dealing with the clinical management of patients with osteoporosis and related disorders, as well contributions from the basic sciences.[3]

The proliferation of national and international organizations was accompanied by frequent conferences that drew participants from many nations. In 1987 the EFFO organized a consensus development conference dealing with the prevention and treatment of osteoporosis. The conference, held in Aalborg, Denmark, had 700 people attendees, including representatives from the medical profession, the pharmaceutical industry, and government ministries. Its report promoted the use of estrogen-progesterone and calcium therapies in the prevention of bone loss and emphasized the importance of exercise. The report also discussed fluoride, vitamin D, anabolic steroids, calcitonin, cyclic treatments with one of the bisphosphonates or calcitonin, the need for research in bone physiology, the necessity of identifying individuals at high risk for developing osteoporosis or fractures, and the importance of crafting preventive and treatment strategies that were inexpensive, effective, and capable of winning patient acceptance.[4]

Three years later another consensus development conference—sponsored by the National Institute of Arthritis and Musculoskeletal and Skin Diseases, the EFFO, and the NOF—convened in Copenhagen. Approximately 2,000 people were in attendance, an indication of the significance now accorded the diagnosis of osteoporosis. The *American Journal of Medicine* published fourteen of the fifteen papers delivered at the conference, plus the final report. Although an international conference, it was clear that Americans occupied a dominant position in osteoporosis research: nine of the fifteen papers were from the United States, two from France, and one each from Australia, Belgium, Great Britain, and Denmark.[5]

These papers provided an overview of the then current state of knowledge about osteoporosis. In his presentation, Charles H. Chestnut III suggested that net bone mass in the aged was a function of peak bone mass at skeletal maturity (probably during adolescence). He endorsed an observation, made by Charles Dent in 1972, that senile osteoporosis was a pediatric disease whose foundation had been laid in childhood and adolescence.[6] Changes in bone mass—which included genetic and environmental factors, although the relative contributions of each remained unknown—either prevented or contributed to fracture.[7] In a discussion of the relationship between exercise and bone density, John A. Eisman and his Australian colleagues noted that although cross-sectional studies demonstrated positive correlations between physical activity and bone density, prospective studies were less clear. These researchers recommended moderate weight-bearing exercise during youth and adulthood, and they concluded that starting such a regimen in postmenopausal and later years was too late. An excess of physical activity, however, had a detrimental impact on both women and men.[8]

Two papers dealt with estrogen therapy. The first, by Robert Lindsay, maintained that this therapy reduced both the rate of bone loss in estrogen-deficient women, regardless of age, and the risk of fractures of the hip and distal radius by about 50 percent. Moreover, prospective data indicated an 80 percent reduction in radiologically evident vertebral deformity.[9] Steven R. Cummings, on the other hand, insisted in his paper that decisions about estrogen therapy had to take into account potential benefits, risks, uncertainties, and the patient's values. Observational studies suggested that this therapy led to a decrease of about 50 percent in cardiovascular disease. There was no immediate upturn in the risk of breast cancer, although long-term estrogen therapy (more than ten years) was associated with an increased diagnosis of such cancers. Nonetheless, Cummings stated, given the protective effects of hormone therapy against cardiovascular disease, its potential benefits outweighed all other effects of this treatment.[10]

Other papers discussed alternative therapies. Jean-Yves Reginster provided data that supported the use of calcitonin as an alternative in preventing and treating bone loss in postmenopausal women.[11] In a discussion on the role of calcium, Heaney noted that an adequate intake would not prevent all osteoporotic fractures, but a low intake would contribute to the occurrence of fractures.[12] Two authors had reservations about the efficacy of fluoride, anabolic steroids, and other therapeutic modalities; one paper was optimistic about the future of bisphosphonate therapy; and another pointed to the absence of well-documented examples demonstrating the reversibility of bone mass loss in secondary osteoporosis. The final papers provided discussions of a variety of subjects, including the diagnosis of osteoporosis, the role of bone mass measurements, and the search for biomarkers.[13]

The conference's final report offered a definition of osteoporosis that quickly became the accepted standard: "a disease characterized by low bone mass, microarchitectural deterioration of bone tissue leading to enhanced bone fragility, and a consequent increase in fracture risk." The report listed a series of risk factors, although it conceded that the understanding of these factors was "incomplete." It endorsed estrogen therapy as the drug of choice after menopause and summarized the various forms of bone mass measurement. The final report concluded by identifying the need for research in seven areas. First, to enhance the epidemiology of osteoporosis. Second, to improve the understanding of how bone remodeling was regulated. Third, to seek ways of reducing postmenopausal, aging-associated bone loss. Fourth, to determine the factor that affected the accumulation of bone tissue during growth and maturation. Fifth, to identify ways of restoring healthy bones in osteoporotic patients. Sixth, to find ways to predict the likelihood of the disease and to monitor its response to therapy. Finally, to expand knowledge about the risk factors and causes of falling among the aged, and to design effective preventive interventions.[14]

In summarizing the deliberations at the symposium, Roger Smith (a consultant physician at an English orthopedic center) noted that misapprehensions abounded. The consensus statement that osteoporosis caused 1,300,000 fractures in the United States alone was questionable. "Fractures have many causes, of which osteoporosis is only one. Had the panel already forgotten about falls?" While cell biologists were trying to understand bone physiology, and epidemiologists were discussing the value of bone mass screening, clinicians were awaiting the development of a truly safe agent that would increase the amount of structurally useful bone and reduce the rate of fractures. Would such an agent be a growth factor or a morphogenetic protein (one related to normal development)? How would it be administered? "Those who deal with an aging population would like these answers quickly."[15]

By the late twentieth century, the ever-increasing proportion of elderly persons in the population, and their coalescence into a distinct group with their own interests, had transformed them into a major political force. Organizations such as the AARP played an important part in representing their concerns. Under these circumstances osteoporosis—a diagnosis that clinicians increasingly identified with the aging process—became even more prominent and was presented as a threat to the health of the elderly.

Reflecting this concern with the health of older Americans, the National Institute on Aging (NIA) convened a Workshop on Aging and Bone Quality in 1992. Its objectives were to review current knowledge and to develop recommendations for further research. Despite decades of studies, many claims about bone physiology were largely "subjects of speculation, awaiting data," noted two members of the Geriatric Program of the NIA. They pointed out that investigations into age-related changes in bone quality could be aided by considerations of analogous lines of aging research in other areas. Heterogeneity was one such example. Perhaps age-related changes in bone quality were not a universal phenomenon, but occurred only in subgroups. Other observations suggested that there was considerable variability in the changes due to aging.[16]

At the opening session of the 1992 conference, A. Michael Parfitt focused on fracture pathogenesis. He pointed out that a decade earlier, he asked the attendees at an NIH consensus conference to consider that bone quality characteristics might contribute to bone fragility. His suggestion fell on deaf ears, but within five years his notion had traveled from heresy to dogma, even in the absence of confirmatory data. During these same years, however, Parfitt had become more skeptical about the practical importance of bone quality. He suggested that participants in the 1992 meeting consider whether there was a role for bone quality. If so, what was its magnitude and what could be done about it?[17]

The relationship between bone quality and fragility factors was extremely important. The prevailing view was that bone density correlated with strength. But was reduced density a sufficient explanation for fragility fractures? If not, what was the role of densitometry in screening, if the results did not predict the risk of fracture? In addressing this issue, Heaney pointed out that virtually every study that contrasted the bone mass density in patients with and without fragility factors found a substantial overlap between the two populations. It was common to find individuals with fractures whose bone mass was within or above the mean of the contrast group. The reverse was also true; some individuals who did not sustain fractures had bone mass well below the mean of those who did. In Heaney's opinion, it was important to look beyond the bone mass density hypothesis.[18] Susan M. Ott added to Heaney's observations when she pointed out that the effect age had on fracture incidence was independent of bone mass. Trauma was such a major factor, she noted, that it was surprising that there were few studies that controlled for it. Ott agreed that bone mass was a risk factor for fracture. Nevertheless, for a 60-year-old woman with a bone mass 2 standard deviations (SD) below the mean, the risk of fracturing a bone during the following year was about 7 percent. Bone failure, she concluded, was "a complex, multifactorial disease." Other structural features that contributed to bone strength could be called "bone quality," and without consideration of such factors, it remained "doubtful that we will ever understand bone failure."[19]

Other papers added to the unanswered questions about bone fragility. One finding—that prevalent (widespread) fractures were strong predictors of future fracture risk—suggested, but did not prove, that there could be other risk factors acting independently of bone mass.[20] The epidemiology of fragility fractures revealed considerable heterogeneity. Bone mineral density had been extensively studied. But hip, spine, and distal forearm fractures revealed variations in their patterns of incidence that were related to age, sex, race, ethnic group, geography, and season. American and African blacks, for example, had lower fracture rates than whites, both in the United States and Africa. The low fracture rate among American blacks reflected their greater bone mass, but this explanation was not true for African blacks. Calcium deficiency among the latter had a negative effect on bone density; their low fracture rates reflected a higher bone turnover, better hip BMD, and sturdier trabeculae, all of which were due to greater physical activity. Yet the reasons for such variations remained unclear.[21]

The papers presented at the 1992 conference and the ensuing discussions revealed that decades of research had failed to illuminate many aspects of bone physiology. Participants themselves identified a series of questions for which they had no answers.

Why was the incidence of hip fractures in communities around the world less than would be expected from bone density levels? Were some aspects of economic development bad for the skeleton (e.g., industrial pollution, a lack of physical activity, trace element deficiency, etc.)? Why did bone remodel in the first place, and what happened to this ability to model as a function of age? What was the molecular basis for the decline in the tensile strength of bone with age? How did microdamage weaken bone?[22]

The following year another international conference, sponsored by the Health Conference on Osteoporosis, met in Paris. The theme of the meeting was "can we make osteoporosis a disease of the past?" The responses to this ostensibly simple question once again revealed the lack of unanimity within the osteoporosis community. To Reinhard Ziegler (a German endocrinologist), the question was provocative. What diseases, among those that were dependent on individual behavior and aging, might disappear? He observed that strategies such as controlling blood pressure or blood cholesterol could lower the incidence of dependent diseases, such as coronary infarctions or strokes. Nevertheless, these entities would continue to exist and cause morbidity and mortality, especially among the aged. Drug treatments such as ERT would reduce the frequency of but not eliminate osteoporotic fractures. Thus, Ziegler noted, unhealthier individuals might live longer because of certain medical strategies, but at the expense of suffering more age-related diseases, such as osteoporosis. Hence the disappearance of osteoporosis was an illusion.[23] A quite different answer came from Gideon A. Rodan (a researcher with the Merck pharmaceutical company), who insisted that increased knowledge about the pathophysiology of bone loss and the ability to control it by HRT or other pharmacological means offered "good hope for making osteoporosis . . . a disease of the past."[24]

Nine of the papers delivered at the 1993 conference focused on peak bone density, bone loss, bone loss measurements, fracture risks, and therapeutic possibilities. There was general agreement that osteoporosis was a serious disease, particularly among older women. Nor was there any overt hostility among conference participants. Yet it was clear that there was little unanimity in determining the etiology of osteoporosis or evaluating therapies for it. Could osteoporosis be explained by events that occurred during the first two decades of life? Did bone loss commence in the premenopausal stage? Could peak bone mass and bone loss explain fracture risks without considering variables such as architectural bone abnormalities, microdamage, geometric properties, and trauma? Could risk assessment be used to develop a cost-effective prevention program and guide therapy? Did bone loss continue through life, or did it decrease in elderly persons?[25]

A variety of answers were given to such questions. Pierre J. Meunier and his colleagues at the Edouard Herriot Hospital in Lyon, France, were optimistic. Preventive measures could be achieved in childhood by increasing bone mass through calcium intake and exercise, by using ERT after menopause, and by using vitamin D and calcium supplements for later prevention in the elderly.[26] Cummings and Michael C. Nevitt had a quite different perspective. They emphasized the role of falls and the lack of research attention paid to traumas that caused fractures. The vast majority of falls did not result in fractures, even among women with low bone mass. The mechanics of falling—including the location of the impact, the energy of the fall, and absorption of that energy—and bone strength were involved in fractures. All of the subjects who suffered a hip fracture fell onto a hip. Those who landed on an outstretched hand or who broke the momentum of their fall by grabbing onto or hitting an object had about one-third the risk of fracturing a hip, compared with women who did not. "The orientation of the fall and site of impact," Cummings and Nevitt concluded, "were by far the most important factors in determining the type of fracture that will result from a fall."[27] Such observations, however suggestive, elicited little interest among clinicians preoccupied with a biomedical model and intent on developing pharmaceutical interventions. If falls, rather than bone loss among the aged, were responsible for hip fractures, then how valid was the diagnostic category of osteoporosis? This question, however, was never asked.

The growing international preoccupation with osteoporosis led the WHO, in collaboration with the EFFO, to form a study group in 1992 to assess fracture risk and its application to osteoporosis screenings for postmenopausal women. The study group ultimately included about seventeen individuals from several European nations, the United States, Australia, and Russia, and it received support from the Rorer Foundation, Sandoz Pharmaceuticals, and SmithKline Beecham. In 1994 the group issued a report that was destined to have considerable influence on medical practice in the United States and elsewhere. Its members noted that the worldwide increase in life expectancy was accompanied by a rise in the prevalence of hip fractures, which added to health care costs. Interest in osteoporosis among experts in bone disease was very high, but general practitioners remained unaware of the problem. The report conceded that the views of bone experts on the assessment and treatment of the disease were "not . . . consistent." Nor had recent consensus development conferences come to any agreement on whom to treat. Hence the group decided to evaluate available methods for assessing fracture risks and their suitability for use in screening for osteoporosis. The absence of randomized controlled trials of screening using bone density measurements was a formidable barrier. Moreover, the feasibility of such studies remained problematic, given the long interval between the onset of bone loss and the occurrence of fractures. Nevertheless, the group decided that its report would have three objectives. First, to define the value and limitations of screening tests for osteoporosis. Second, to identify the population that was at risk. Finally, to determine the costs and benefits of screening for low bone density.[28]

It was hardly surprising that screening would become central to the identification of osteoporosis. By this time, Americans were being urged to have mammograms for breast cancer, pap smears for cervical cancer, PSA (prostate-specific antigen) tests for prostate cancer, and blood pressure evaluations for hypertension, to cite only a few such examples. Implicit in these tests was the assumption that early identification of these conditions and appropriate interventions could prevent serious morbidity or mortality. Osteoporosis researchers and clinicians were well aware of the presumed benefits of screening, and they were interested in developing appropriate standards and numerical scales that could guide practitioners. Their efforts also received financial support from pharmaceutical firms seeking to expand the market for their products and from manufacturers of screening devices. Screening was especially compatible with the belief that diseases could be prevented or conquered by innovation and that ever-increasing investments in health would yield high returns.

Screening for osteoporosis had always presented problems, if only because of the absence of clear and unambiguous numerical categories. In its 1994 report, the WHO study group identified four general diagnostic categories for postmenopausal osteoporosis. The norm was a young, healthy, adult woman. The first (normal) category was when BMD was within 1 SD of the reference group. The second was low bone mass (osteopenia), which was more than 1 SD but less than 2.5. The third was osteoporosis, which was 2.5 SD or more. The final category was severe osteoporosis, which was more than 2.5 SD and involved one or more fragility fractures. The report conceded that its cutoff values were "somewhat arbitrary," but it pointed out that a measured value of 2.5 SD below the mean for healthy adult women identified 30 percent of all postmenopausal women as having osteoporosis.[29] Ultimately this classification system became a T-score, which was a number that quantified the BMD of a person relative to the average among white women age 20–29 (a T-score above –1). Since virtually all postmenopausal women lost some bone mass, they had negative T-scores.

To diagnose osteoporosis by a numerical scale was by no means unique. In the last third of the twentieth century, numbers and scales began to define disease. Prior to World War II, blood pressure readings of 120/80 were regarded as normal, and it was assumed that such readings rose with age. A half century later, hypertension was redefined in terms of a broader syndrome, and a reading of 140/90 became the threshold for treatment. Similarly, the figures indicating desirable cholesterol and glucose levels have been constantly lowered. Shifting numbers, of course, result in a dramatic expansion of the population requiring treatment. In the case of hypertension, there is clear evidence that a high reading is dangerous, although there is no agreement on the optimum treatment threshold. Osteoporosis presented a quite different case, if only because the evidence that a low BMD (as expressed by the T-score) was a good predictor of fracture was, at best, equivocal.[30] The data on fractures often revealed that the relationship between low or high BMD and a fracture was problematic. The emphasis on numerical scales nonetheless conveyed an aura of certainty; numbers were presumably objective. Moreover, numbers seemed to provide a scientific basis for making medical decisions, rather than relying on the clinical judgment of individual practitioners.

Was it possible to maximize peak bone mass in early life? Could adequate nutrition, exercise, and avoidance of smoking achieve this goal? A healthy lifestyle was obviously desirable. But the WHO group conceded that there was little evidence to support the claim that population-based strategies could reduce the incidence of fractures by modifying poor early-lifestyle behaviors. Knowledge about medicine's ability to modify peak bone mass was also limited. The most feasible way to prevent fractures due to osteoporosis was to prevent bone loss in those identified as being at high risk. The WHO report reached several conclusions. A technology to assess low BMD with sufficient specificity and sensitivity was available (e.g., SPA and DPA). Similarly, HRT was an effective method of reducing bone loss. Nevertheless, the argument that *all* women should be screened was a poor one. Selective screening for women within five years of menopause was worthwhile, in order to stratify risk and offer such proven interventions as HRT. Women on long-term HRT should be exempted from screening, given the effectiveness of this therapy in preventing bone loss. There was also a good case for screening women in older age groups. An optimal age was 65, when the risk of hip fractures was still low, the effects of intervention had been proven, and the cost-effectiveness of screening was favorable.[31]

In an article summarizing the WHO report, John A. Kanis (a member of the study group) insisted that osteoporosis, whose natural history was "adequately understood," was an important social problem. Screening technology was "simple and safe," sensitive, specific, and acceptable to people. HRT was a proven and effective treatment, and the target population was clearly defined. Facilities for the diagnosis and treatment of osteoporosis were cost-effective. "The principles for screening," Kanis concluded, "are to a large extent met in relation to osteoporosis."[32]

Bone mineral measurements, according to the NOF's Scientific Advisory Board and the Board of National Societies of the EFFO, were comparable to the use of blood pressure readings to assess the risk of strokes. Blood pressure was a physical characteristic distributed throughout the population, as was bone density. Just as patients above a cutoff level for blood pressure were diagnosed as hypertensive, patients below a bone mineral cutoff level were given a diagnosis of osteoporosis. The ability of BMD measurements to predict fractures was as good as the ability of blood pressure readings to predict strokes. The group conceded that normal BMD was no guarantee that fractures would not occur. But if BMD was in the osteoporotic range, fractures were likely.[33]

A year after the publication of its report on fracture risks and BMD screening, the WHO's Division of Drug Management and Policies created a working group, composed of drug regulators and scientists, to develop principles and methods to guide clinical trials dealing with the efficacy and safety of drug treatments for osteoporosis. The guidelines had to consist of evidence-based principles, rather than detailed prescriptions for the conduct of such studies; they had to be sufficiently broad to encompass all types of osteoporosis, as well as their prevention and treatment (including non-pharmacological interventions); and they had to be relevant to all those involved in the development and evaluation of interventions, including the pharmaceutical industry, drug regulatory authorities, and individual scientists.[34] While there were many similarities between these guidelines and those issued by the FDA in 1994 for the evaluation of various agents used in the treatment of postmenopausal osteoporosis, the WHO guidelines placed far greater emphasis on the impact of the pre-clinical results than on the clinical evaluation process.[35]

The WHO's emphasis on the establishment of guidelines for preclinical evaluations and clinical trials reflected a conviction that in the future, drugs would play the dominant role in treating patients already suffering from or at risk for osteoporosis. It also represented a shift in research strategies. During and after the late 1960s, the focus was on in vitro systems, with the widespread use of tissue and cell culture techniques. In the 1990s, there was an increasing use of animal experimentation to deal with fundamental problems involved in the pathogenesis of bone diseases such as osteoporosis. Technical advances led to the development of several animal models of osteoporosis, which presumably would produce a better understanding of the disease in humans and enhance the ability of researchers to test antiosteoporotic therapies.[36]

Preclinical studies related to osteoporosis had several objectives. First, to establish the relationship between the effects of a drug on bone mass and bone strength. Second, to understand the mechanism of the drug's action. Third, to demonstrate its results in animal models. Fourth, to establish the impacts of long-term exposure. Finally, to examine the influence of that particular intervention on fracture repair. When preclinical studies (phases I and II) resolved such issues, investigators could move into phase III trials to test the drug's efficacy on bone mass / mineral density. The effectiveness of the drug would be established if the results in humans were similar to those observed in animal models. Phase IV (post-marketing surveillance) would provide data on the influence of the drug on fracture rates.[37]

This research strategy was designed to overcome the problem of following large numbers of individuals who were at risk for fracture over long periods of time, in order to demonstrate a drug's efficacy. The results of the preclinical evaluations would determine the endpoints (specific conditions which, when reached, would end a clinical trial) required for the different phases of the drug undergoing clinical development. If, after the administration of this drug, the quality of the bone was normal and the mechanism of the drug's action was understood, results that produced a normal BMD might be an appropriate endpoint. If the quality of the bone was uncertain but the mechanism of action was understood, the point at which likely potential fractures were avoided would be an appropriate endpoint.[38] The guidelines, which were drawn up in collaboration with pharmaceutical-industry scientists, clearly enhanced the central role of drugs in the prevention and treatment of osteoporosis.

The WHO took on an even broader role when it launched an international osteoporosis education project, the cornerstone of which was the preparation of a master document on the management and prevention of this condition. "Osteoporosis," noted its Task Force for Osteoporosis, "was not classified as a disease until relatively recently; previously it was considered an inevitable accompaniment of aging." Now it was recognized as "a progressive systemic disease," characterized by low bone density and microarchitectural deterioration of bone tissue, with a consequent increase in bone fragility and susceptibility to fracture. The task force supported a wide-ranging campaign designed to educate the general population, physicians, and health authorities, and its members offered recommendations for further research. If there was little new in this report, it nevertheless contributed to the belief that osteoporosis was a global health problem that was common in developed countries and likely to become so in developing countries as longevity in their populations increased.[39]

The WHO's diagnostic criteria for low bone mass and osteoporosis, although widely accepted, was not immune from criticism. The Study of Osteoporotic Fractures Research Group at the University of California, San Francisco, evaluated nearly 10,000 white women over the age of 65, and these investigators concluded that decision making based on T-scores was not justified. Across manufacturers of densitometers, comparable BMD values yielded different T-scores. Moreover, reference peaks, means, and standard deviations were difficult to determine. "We propose," the research group concluded, "that the field move toward risk-based BMD cut points, eliminating the need for peak BMD, reference means, and standard deviations . . . and freeing us from the tyranny of the T-score."[40]

The harshest criticism came from Richard D. Wasnich of the Hawaii Osteoporosis Foundation. He noted that consensus involved compromise between opposing viewpoints. When there was substantial agreement, consensus guidelines could be useful. But hazards were by no means absent. If one viewpoint was right and the other wrong, a compromise would move away from the truth. Hence when true consensus did not exist, both sides of the debate should be heard. Wasnich was especially critical of recent consensus statements, which emphasized *how* to treat osteoporosis. The equally important question of *who* to treat was ignored. Bone density, Wasnich insisted, was a risk factor, not a diagnostic test. Thus making a diagnosis based on a single risk factor, at a single point in time, was tenuous. After all, he argued, there were other risk factors. Moreover, low bone density was sometimes a manifestation of diseases other than osteoporosis, such as hyperparathyroidism or renal hypercalciuria (excess calcium in the urine), and these had to be excluded by appropriate laboratory evaluations. What physicians needed to know, Wasnich stated, was not how their patient compared with a "normal" reference value derived from a young female population. Rather, they needed an estimate of the *absolute* fracture rate, which could be derived directly from the bone density value. The debate, Wasnich concluded, ought to be continued and not stifled by "expert" consensus statements.[41]

The WHO model of osteoporosis, three researchers pointed out, assumed that bone fragility depended only on a single mean value of BMD. Such a model was "overly simplistic," because it lumped all patients together within arbitrarily defined ranges. It ignored confounding factors—such as variability in equipment; in the age, gender, or ethnicity of patients; and in skeletal site—all of which contributed to disparate results. Moreover, mean BMD, as a single number, could not distinguish patients with fractures from healthy individuals. Based on their study, these authors concluded that fracture risk had to include quantitative characteristics of the BMD distribution that represented macro-structural information about the spine.[42]

Even John A. Kanis (a member of the WHO study group that developed the WHO guidelines) and his associates found that the use of a threshold for osteopenia (a T-score of –1.0) was associated with only modest increases in hip fracture risk, compared with that of the general population. The fracture risk among post-menopausal women at age 50 who were at the threshold for os-teopenia was less than that of the general population at that same age. This was due to the choice of age range (20–29) for the refer-ence value of BMD on which the T-score was derived. Although a T-score of –2.5 or lower was a reasonable treatment threshold, it was not accurate enough to predict absolute risk, which varied by age. Kanis concluded that it was important to derive absolute (e.g., ten-year) risks to stratify the risk factor for individuals and thus direct interventions.[43]

Such reservations notwithstanding, the T-score quickly be-came the means by which osteoporosis was diagnosed. Clinicians in private practice were generally unaware of the differing points of view and debates among osteoporosis researchers. Moreover, much of these physicians' information came from pharmaceutical firms eager to find a market for the increasing number of drugs that purportedly prevented osteoporosis and from manufacturers of imaging equipment.

EPIDEMIOLOGICAL STUDIES

The rising amount of publicity concerning osteoporosis, the founding of international osteoporosis organizations, and the ac-tivities of the WHO were accompanied by an acceleration of in-terest in broad epidemiological studies that transcended national boundaries. During and after the 1980s, these studies, which were designed to identify fracture rates and risk factors, not only con-firmed but also strengthened the legitimacy of osteoporosis as a disease, rather than as a product of the aging process. Substantial geographic differences in the incidence of hip fractures, and large differences in sex ratios for hip fractures in European countries, were already known. The hope was that comparative epidemiological studies might lead to the identification of hitherto unknown risk factors and thus provide clues for fracture-prevention programs. Even if many risk factors could neither be prevented nor modified (e.g., family history, past fractures, loss of vision), epidemiological studies could identify risk groups amenable to drug treatments or to such preventive measures as the use of protective hip pads or changes in their home environments.[44]

In 1986 six nations (Portugal, Spain, France, Italy, Greece, and Turkey) launched the Mediterranean Osteoporosis Study (MEDOS), with the cooperation of the WHO and the EFFO. This study involved fourteen centers and had three components. The first consisted of collecting existing register data on hip fractures from the thirty-one European ministries of health. The second was a study of the hip fracture incidence of over-50-year-olds in the participating centers within these Mediterranean countries. The third was a case-controlled study comparing 2,816 cases and 5,369 controls. The goal was to gain information on the incidence of, risk factors for, and means of preventing hip fractures.[45]

The MEDOS study of women age 50 or older, conducted over a one-year period, identified a low body mass index (BMI), short periods of fertility, a low level of physical activity, a lack of exposure to sunlight, limited milk consumption, no consumption of tea, and a poor mental score as "significant" risk factors. A moderate intake of alcohol was a protective factor in young adulthood, whereas there was no evidence that smoking posed a significant risk. The MEDOS group conceded that the total amount of attributable risk was high, but it did not follow that identification of such risk factors in the community would be a useful screening tool. The group noted that studies of risk factors did not demonstrate causality, even when the associations were plausible. Moreover, a composite score from the independent risk factors revealed its relatively low sensitivity and specificity. On the other hand, the sensitivity and specificity of the composite scores were comparable with those for BMD in discriminating between individuals with and without osteoporotic fractures. Thus screening for risk factors prospectively was worthy of further study.[46]

In the United States, Bernadine Healy (director of the NIH) launched the Women's Health Initiative (WHI) in 1991. It was one of the largest efforts to assess risk factors that affected the health of older women. The WHI contained three major components. Its centerpiece was a huge clinical trial involving 57,000 postmenopausal women recruited to test the effects of low-fat diets, HRT, and vitamin D and calcium supplements on heart disease, cancer, and osteoporosis. The other components included an observational study to follow several hundred thousand women over time, and a nationwide community intervention and prevention study. The study was expected to take a decade to complete and cost more than $600 million. Healy came under considerable criticism, especially from other women who noted that the study was dominated by male investigators. Healy was not especially sympathetic to earlier feminists, many of whom argued that there were no differences between the sexes. Instead, Healey believed that there were significant biological differences between women and men, and she insisted that physicians and medical scientists, irrespective of their gender, were best suited to study matters pertaining to women's health. The results of the WHI initiative, as we shall see in chapter 7, had important consequences.[47]

Unlike the WHI undertaking, international epidemiological studies focused exclusively on osteoporosis. The European Prospective Osteoporosis Study (EPOS) attempted to identify the causes of the epidemic of hip and spine fractures affecting Europe and most other countries. It grew out of two earlier concerted actions by the European Union. The European Vertebral Osteoporosis Study (EVOS) attempted to determine the prevalence of radiographically defined vertebral deformity as a marker for vertebral osteoporosis. Thirty-six centers in nineteen European countries participated. EVOS found a three-fold variation in the occurrence of deformity across Europe, a fact, it was hoped, that would, in the future, establish the causes of such variation and thus lead to the development of population-wide preventive studies. The second study, done under the auspices of COMAC-BME (Comité d'Actions Concertés-BioMedical Engineering), was designed to standardize bone density results obtained from dual-energy x-ray absorptiometry (DXA) machines made by different manufacturers.[48]

EPOS was by far the largest study of spine, hip, and peripheral fracture rates. It commenced in 1993 with a questionnaire to EVOS cohorts, and continued with a yearly follow-up. Some of the results were surprising. About 12 percent of both men and women had vertebral deformities (using a 3 SD cutoff). Women had lower rates at age 50, and higher rates at age 75, than men. An analysis of risk factors suggested that in women, a moderate level of physical activity protected against vertebral deformity. In men, the highest level of physical activity, usually an occupational one, was a positive risk factor.[49]

As the study continued, other findings contradicted long-held convictions. Despite the belief that osteoporosis was largely a female disorder, the data confirmed its frequent occurrence in men.[50] Lifestyle factors—smoking, alcohol intake, level of physical activity, and milk consumption—showed no consistent association with incident (localized) vertebral fractures. These findings were in stark contrast with those of the MEDOS study. The data, EPOS members conceded, "suggest that modification of lifestyle risk factors is unlikely to have a major impact on the population occurrence of vertebral fractures. The important biological mechanisms underlying vertebral fracture need to be explored using new investigational strategies."[51] Even more surprising was the fact that BMD was less important in explaining variations in the incidence of upper limb fractures in women across diverse European populations, compared with the effect of other factors. These included a location-specific risk of falling; a personal/family history of fractures; or factors that were associated with the risk of falling, such as the amount of time spent walking or cycling.[52] Equally puzzling was the fact that there were significant variations in the occurrence of hip, distal forearm, and humerus fractures across Europe, with higher incidence rates in Scandinavia than in other European regions.[53]

The largest epidemiological study of osteoporosis in the United States began in 1997, with the creation of the National Osteoporosis Risk Assessment (NORA). This study was managed and administered by the Merck pharmaceutical company, in collaboration with the International Society for Clinical Densitometry. A steering committee, composed of osteoporosis researchers internal and external to the sponsoring organizations, had oversight authority. NORA was initially designed to assess the association between osteoporosis risk factors and low BMD, and to examine the relationship between BMD, other risk factors, and a short-term (one-year) incidence of fractures. In NORA, over two hundred thousand postmenopausal women underwent peripheral bone densitometry or ultrasonography of their heel, finger, or forearm in physicians' offices. They also completed questionnaires assessing risk factors and, approximately a year later, documenting new skeletal fractures. NORA included a registry that would permit an estimation of the relationship between known or suspected fractures and fracture risk. A primary-care education and awareness program was also part of the study.[54]

Serious issues were raised because the program was managed and funded by Merck, and involved the International Society for Clinical Densitometry. The potential for bias was by no means absent. Documentation of an increased number of women at risk for osteoporosis provided an ever-larger market for therapeutic intervention. Merck was already marketing an antiosteoporotic drug to compete with Premarin and other estrogen products. Those involved in imaging also stood to benefit as the population for screening expanded. Defenders of NORA, however, felt that both full disclose of the study's funding source and full access to all the data and the data assessment by investigator-authors mitigated against bias. Nevertheless, the ensuing estimates of high prevalence rates for osteoporosis, and the claim that the vast majority of individuals with osteoporosis had never been diagnosed or treated, remained as disquieting elements.[55]

NORA's longitudinal observational study (from September 1997 to March 1999) identified 200,160 ambulatory postmenopausal women age 50 or older, with no previous osteoporosis diagnosis, who were drawn from 4,236 primary-care sites in thirty-four states. Using WHO criteria, 39.6 percent of these women had osteopenia (T-scores between −1 and −2.49) and 7.2 percent had osteoporosis (T-scores of −2.5 or below). Age, a personal or family history of fracture, Asian or Hispanic heritage, smoking, and cortisone use were associated with a significantly increased likelihood of osteoporosis, whereas a higher BMI, African American heritage, estrogen or diuretic use, exercise, and moderate alcohol consumption significantly decreased that likelihood. Low BMD was the single best predictor of a fracture risk in this population. Osteoporosis was associated with a fracture rate that was approximately four times that of an individual with normal BMD. In other words, half of those women in the study had a previously undetected low BMD, which was highly predictive of fracture risk. "Given the economic and social costs of osteoporotic fractures," concluded the study's authors, "strategies to identify and manage osteoporosis in the primary-care setting need to be established and implemented."[56] In an accompanying editorial in *JAMA*, Chestnut noted that "NORA confirms what many clinicians and osteoporosis researchers have long suspected, i.e., that a significant number of postmenopausal women in primary-care practices have clinically significant low BMD and that such women have an increased risk of incident fractures within 1 year." Given the availability of at least five FDA approved therapies for prevention and treatment, Chestnut observed, such underidentification was "unfortunate."[57] The finding that low BMD was a fracture risk was confirmed by another NORA study in 2002.[58] In 2004 an additional analysis showed that low BMD in postmenopausal women 50–64 years old showed a relative risk of fractures similar to that found in women 65 years of age and older.[59]

If most fractures were the result of falls, was it possible to isolate risk factors that independently predicted a significantly increased risk of falling? At the Seventh International Symposium on Osteoporosis, a group of investigators used data on 66,134 postmenopausal women living in the community and enrolled in NORA and identified eighteen risk factors. The largest predictor of a fall risk was a history of falls. A fall in year one predicted future falls in years three or six. The authors' multivariate analysis found seventeen additional risk factors that were associated with incident falls (those causing physical damage), although these were less important. Women with less education were not as likely to report a fall as women with a college degree. Older age was associated with a risk of falling, particularly among those over 80, compared with those who were 50–69. Various conditions— depression, strokes, limited physical functioning, prior fractures, a BMI greater than 30, kidney-liver disease, poor memory, poor health, poor hearing, diabetes, hypothyroidism, a loss of height from peak height, and no history of estrogen therapy—predicted a risk of falls. Falling increased linearly among those with multiple risk factors. The investigators called for more studies to validate the predictive value of these risk factors, in order to assess their potential to identify candidates for fall-prevention programs.[60]

The NORA study of risk factors for falling was somewhat idiosyncratic, in that there was no mention of the relationship of these risk factors to the diagnostic category of osteoporosis. The one that was linked with osteoporosis—no history of estrogen therapy—had the lowest odds ratio (1.09, compared with the highest of 2.67) out of the total of eighteen risk factors.[61] If falling among postmenopausal women was responsible for fractures, what was its relationship to osteoporosis? The risk factors for falling, after all, largely involved conditions normally associated with the aging process. "The liability of old people to tumble and often to injure themselves is such a commonplace of experience that it has been tacitly accepted as an inevitable aspect of aging," wrote a British physician in 1960. In studying 500 falls, he concluded that different modes of falls reflected a defect in a person's control of posture and gait, which followed the age-related change of declining numbers of nerve cells in the brain stem, cerebellum, and other centers of motion.[62] Many studies questioned the etiologic importance of osteoporosis in age-related fractures. John M. Aitken noted that logical reasons may have led to an assumption that there was a cause and effect between osteoporosis and femoral-neck fractures. Yet 98 percent of these fractures resulted from a fall, and their causal relationship to osteoporosis "may be much smaller than assumed hitherto."[63] Cummings found that those with hip fractures were not more osteoporotic than other persons of a similar age, and he called attention to the importance of falling.[64] Melton was critical of such claims and insisted that epidemiological data demonstrated that the risk of fracture varied with BMD at the fracture site.[65]

NORA was followed by several other international epidemiological studies. The Global Longitudinal Study of Osteoporosis in Women (GLOW) was a prospective cohort study involving 723 physicians and over 60,000 women subjects age 55 and older in ten countries. It was an observational longitudinal study designed to improve an understanding of international patterns of susceptibility, recognition, management, and outcomes of care for women at risk for fragility fractures. Over a two-year period (between 2006 and 2008), questionnaires were mailed to women whose names were provided by their primary-care physicians.[66] By this time the WHO had developed its Fracture Risk Assessment Tool (FRAX), which will be discussed further in chapter 7. The risk factors included in FRAX were age, sex, BMI, a personal history of fractures, a parental history of hip fractures, current smoking, alcohol intake, glucocorticoid use, and rheumatoid arthritis. Using this instrument, one study found that there was a consistent underappreciation of personal risk factors for osteoporosis and fractures, despite the availability of tools for diagnosis and risk assessment and the existence of what were believed to be safe and effective therapies. The failure of women to appreciate their own risk in sustaining a fracture was a serious barrier to appropriate assessment and management, and the authors of this study called for improved education for both physicians and postmenopausal women concerning these risk factors.[67] Another study challenged the widespread belief that obesity protected against fractures. Its findings instead suggested that obesity was a risk factor for certain fractures, notably those of the ankle and upper leg.[68]

Epidemiological studies such as MEDOS, EPOS, NORA, and GLOW—all of which were international in scope—were by no means the only ones of their kind. Others were confined within individual nations. In Italy, for example, the Incidence and Characterization of Inadequate Clinical Responders in Osteoporosis study found that the incidence of fractures during treatment with antiresorptive agents was considerably higher than those observed in RCTs. The same group of researchers offered two quite different explanations for this result. In 2006 they concluded that inadequate compliance with the treatment and a lack of calcium and vitamin D supplementation were the major determinants of the poor response to treatment. Two years later they suggested that the antifracture efficacy of antiresorptive agents might be lower in real life than what was observed in RCTs and that the benefits tended to wear off after a few years of treatment.[69]

A second Italian study (the Break Study) enrolled 1,249 women age 60 and over who were seeking medical care for a hip fracture. It found that age, smoking, one's family history, and treatments with antiplatelets, anticoagulants, and anticonvulsants were significant predictors of a positive history of hip fractures. The investigators conceded that the study had limitations, however. There was no assessment of the patients' mental status, which jeopardized the reliability of the collected data. In addition, since it was a cross-sectional study, there could be no cause-effect association between risk factors and previous fractures reported by the patients.[70]

The activities of national and international organizations and the numerous epidemiological studies that were undertaken by them had a major impact. They focused attention on the diagnosis of osteoporosis in ways that permitted no doubts about the existence of this condition. The creation of the T-score by the WHO—an organization held in high repute—seemed to provide a measurement that not only was legitimized by the findings of medical science, but also provided a sure guide to therapeutic interventions. By the beginning of the twenty-first century, the claims and warnings from the osteoporosis community had begun to influence primary-care practitioners. The emphasis on screening found that millions of people—both female and male—were at risk for fractures, a finding that resulted in a vast expansion of the numbers of people undergoing drug therapy.

Yet what was largely ignored was the striking lack of unanimity among osteoporosis investigators. Their differences were rarely debated in public, although a careful reading of the medical literature reveals that many of the claims about risk factors and appropriate therapies represented assertions that frequently lacked empirical evidence. It was clear that vitamin D and calcium played important roles in bone development, but it was never proven that taking supplements of both would serve a therapeutic purpose. Long-term randomized controlled trials of fracture vulnerability posed formidable problems, but their absence vitiated many of the claims prevalent in the osteoporosis community. Moreover, the increasingly important role played by pharmaceutical firms in supporting epidemiological studies, individual investigators, and organizations such as the NOF and others raised serious questions about the alleged impartiality of all who were involved.

CHAPTER SIX

Therapeutic Expansion

Although concern with osteoporosis accelerated during and after the 1980s, the therapeutic armamentarium remained relatively limited. ERT was by far the most frequently prescribed medication, even though there were disagreements on how it should be used. Other therapies, including calcitonin, anabolic steroids, sodium fluoride, and human parathyroid hormone, did not compare with ERT. Vitamin D and calcium supplementation were not controversial regimens, and patients were often urged to use them. But the preponderant view was that none of these interventions could compare with the effectiveness of estrogen.

As long as the boundaries of the diagnosis of osteoporosis were circumscribed, drug companies had little incentive to develop alternative therapies and challenge Ayerst's Premarin. But as the numbers of women diagnosed as either osteopenic or osteoporotic grew, interest in developing new therapies increased accordingly, particularly as doubts about the safety of ERT were raised. During and after the 1990s, a variety of new drugs entered the medical armamentarium. They were accompanied by recommendations from specialty groups and organizations stating that virtually all women should consider pharmacological therapy to prevent hip and vertebral fractures. The pharmaceutical industry played a major role through its financial support of both research and clinical trials, to say nothing of publicity campaigns designed to persuade clinicians and patients of the importance and effectiveness of pharmacological therapy to prevent or treat osteopenia and osteoporosis.

ESTROGEN

By the 1980s Premarin—one of the five best-selling prescription drugs in the United States—was the leading drug prescribed for the prevention and treatment of osteoporosis. Its competitors lagged far behind, partly because the Ayerst pharmaceutical company had been promoting it in enthusiastic terms as an antidote for both osteoporosis and menopause. The only other competitor was calcitonin, a drug made by two European pharmaceutical firms. Calcitonin, however, had to be injected every day. Aside from its inconvenience and discomfort, its daily cost ($6 and $7) was beyond the means of many people.

Yet the debate about the health benefits and risks of ERT had never been completely resolved, and uncertainty concerning the public health implications of long-term exposure to supplemental estrogen persisted.[1] The WHI was just getting started, and results from that study were still some years away. Nevertheless, it was clear that there was no agreement about the risks of long-term ERT. Uncertainty, if not confusion, characterized recommendations concerning its use.

In 1992 the American College of Physicians published guidelines designed to counsel postmenopausal women about preventive hormone therapy. Its recommendations were based on a review of the medical and scientific literature since 1970 dealing with the use of long-term hormone therapy to prevent disease or prolong life. Their review of 265 studies summarized the effect of estrogen-only therapy and estrogen therapy plus progestin on endometrial cancer, breast cancer, coronary heart disease, osteoporosis, and strokes. It found evidence that ERT decreased the risk of coronary heart disease and hip fractures. Long-term therapy, on the other hand, increased the risk of endometrial cancer and may have been associated with a small increase in the risk of breast cancer. The heightened risk in the former could probably be avoided by adding a progestin to the estrogen regimen for women with an intact uterus, but the effect of combination hormones on the risks of contracting other diseases was unknown.[2] Based on their literature survey, the American College of Physicians could only come to tentative recommendations. It suggested that all women, regardless of race, consider hormone therapy. Women who had a hysterectomy and those with or at high risk for coronary heart disease would benefit from such therapy. "For other women," the guidelines stated, "the best course of action is unclear."[3]

In many respects such guidelines were a mirror image of the divisions in the osteoporosis community. The American College of Physicians noted that estimates of risks and benefits came from observational studies, which could not provide conclusions about cause and effect. Randomized trials were required in order to prove effects and define their magnitude. Such trials, however, needed to be very large and would have to continue for many years, an undertaking that faced nearly insuperable obstacles. Hence the organization's recommendations were based on the best available current data, but they were subject to possible change as the results of randomized trials became available.

The document also suggested that a "woman should understand the probable risks and benefits of hormone therapy, decide how valuable she considers each of the potential effects of therapy, and participate with her physician whether to take preventive hormone therapy." But how could women and their physicians decide whether or not to begin an ERT regimen, given the disagreements within the osteoporosis community? The specific recommendations of the American College of Physicians, therefore, were at best equivocal and at worst confusing.[4]

Differences over the risk-benefit ratio for ERT were evident in another meta-analysis of the literature in 1993, authored by a group from the Centers for Disease Control and Prevention (CDC), on the topic of breast-cancer risk and the duration of estrogen use. One of the major problems in epidemiological studies

is their heterogeneity. In this case, study design, community versus hospital controls, varying case-control and follow-up designs,
the use of different hormone preparations, and dose-response
slopes all added to the difficulties of coming to definitive answers.
The authors estimated that the risk of breast cancer after ten years
of supplemental estrogen use increased between 15 and 29 percent. "We do believe," they concluded, "that the consistency and
biologic plausibility of our results are cause to invest in basic and
epidemiologic research to answer the question: does long-term
estrogen use increase risk of breast cancer?"[5]

A 1995 study of the relationship between ERT and the risk
of fatal colon cancer in a prospective cohort of postmenopausal
women came to a quite different conclusion. Here the investigation was under the auspices of the American Cancer Society,
which in 1982 had launched Cancer Prevention Study II, a prospective mortality study of roughly 1.2 million Americans whose
average age was 56. By 1989, after seven years of follow-up, 93.2
percent of them were still alive and 6.5 percent had died. At the
end of this follow-up period, 897 colon cancer deaths were observed in a cohort of 422,373 postmenopausal women who were
cancer free at the beginning of the study. The investigators found
that ERT use, whether recent or long term, was associated with
a substantial decrease in the risk of colon cancer. This association
was not altered by multivariate analyses controlling for other risk
factors. The investigators conceded that the limited number of
studies on ERT and colon cancer risk meant that their findings
should be interpreted cautiously, but they also added that the results were sufficiently important to merit further investigation.[6]

Recommendations for and against ERT were common. Members from the Study of Osteoporotic Fractures Research Group
undertook a prospective study of 9,704 women age 65 or older.
Its authors acknowledged that there were few data on the effects
of long-term estrogen use. Nevertheless, this research group's data
supported the finding of earlier case-control and prospective studies, where estrogen use initiated soon after menopause and continued indefinitely was associated with the lowest risk of fracture.[7]

Another study dealing with the effects of estrogen or estrogen-progestin regimens on heart disease also came to positive conclusions: estrogen was linked with improved lipoprotein (HDL and LDL cholesterol) readings and lower fibrinogen levels (a factor to autoimmune disorders, some types of arthritis, and certain kinds of inflammatory diseases), and it provided cardioprotection for women.[8] In a laudatory accompanying editorial in *JAMA*, Bernadine Healy insisted that it was imperative to launch a controlled clinical trial to investigate whether differences in the studied surrogate markers would translate, in time, to less cardiovascular disease, which was the leading cause of mortality among women.[9]

A group of Harvard Medical School researchers, concerned with the risk of breast cancer from ERT, were far less enthusiastic. They analyzed data from the Nurses' Health Study, which had begun in 1976 and was expanded in 1989 to include information from 238,000 nurse participants. During 725,550 person-years of follow-up, the data revealed 1,935 cases of invasive breast cancer. The Harvard investigators concluded that postmenopausal women taking estrogen alone or estrogen plus progestin, compared with those who had never used hormones following menopause, had a higher risk of breast cancer. The increased risk associated with five or more years of hormone therapy was greater among older women (60–64). Women over 55 years of age, therefore, had to consider the risks and benefits of ERT, especially if they had been using this therapy for five or more years. Moreover, short-term estrogen therapy—for up to seven years—in the decade after menopause could not be expected to protect against osteoporotic fractures many years later. The article's publication in the *New England Journal of Medicine* gave it added credence.[10] In an accompanying editorial in the same issue, a Johns Hopkins oncologist wrote that "our fragmented knowledge of therapy with estrogen alone or with estrogen plus a progestin mandated a prospective assessment of the net health effects of postmenopausal hormone therapy. . . . All postmenopausal women should be apprised of our current understanding of the risks and benefits of hormone-replacement therapy."[11]

At the same time that the Harvard Medical School study dealing with ERT and breast cancer appeared, the Office of Technology Assessment of the U.S. Congress issued a report on the
effectiveness and costs of osteoporosis screening and of HRT. In
general the report, which included an analysis of the benefits and
risks of such therapy, was favorable. Based on an analysis of the
literature, the Office of Technology Assessment found that HRT
was more protective against fractures if it was begun within five
years of menopause and was used for longer than ten years. The
evidence dealing with a link between estrogen therapy and breast
cancer was inconclusive, the report stated, largely because case-
control and cohort studies could not control for biases and confounding factors. The inconsistency of the results suggested that
the effect of HRT on breast cancer "is likely to be small." The
agency's report found that estrogen use alone increased the odds
of endometrial cancer, but the addition of a progestin provided a
protective antidote. ERT also resulted in a reduced risk of heart
disease. The study's economic analysis indicated that a life-long
course of ERT for all women at menopause was a good investment for society.[12]

How did women react in the face of contradictory evidence
about risks and the various therapeutic recommendations? The
answer to this question is extraordinarily complex. A literature
review (in 2005) of 112 papers published between 1980 and 2002
dealing with women's perceptions about the risks and benefits
of HRT found both positive and negative beliefs. Interest in its
helpful aspects was often countered by concerns over its potential
adverse effects. The use of HRT was also strongly influenced by
short-term symptomatic considerations, rather than knowledge
about its long-term benefits. Moreover, many women believed
that menopause was a natural event, with no treatment required.[13]
They also did not link fractures with bone health. On the contrary, these women attributed fractures to particular situations,
such as slipping on ice or tripping on uneven ground. Fractures,
in other words, were due to accidents and thus not a cause for
personal alarm.[14]

Several microstudies provided more detailed information. In a survey that polled 7,667 white women age 65 or older who participated in the Multicenter Study of Osteoporotic Fractures, the investigators found that about 17 percent were HRT users, 27 percent were past users, and 55 percent were non-users. The self-reported primary reasons for current users to have initiated this therapy included a hysterectomy (43.5%), menopausal symptoms (39.3%), treatment prescribed by a physician (38.7%), and the prevention or treatment of osteoporosis (33.6%). Among the former users of estrogen, the main reasons for starting therapy included treatment prescribed by a physician (44.7%), menopausal symptoms (49.2%), and a hysterectomy (28.5%). About 30 percent of these past estrogen users reported that they did not need it, and 16.4 percent stopped because of undesirable side effects, with bleeding being the most common (45%). Those who never began estrogen therapy often reported that they feared the medication was harmful (38.1%) or felt they did not need it (29.5%). The investigators noted that despite the publicity given to estrogen therapy in the prevention of osteoporosis and in reduced rates of cardiovascular disease, a large proportion of older women remained skeptical.[15]

Another survey of 1,082 women age 50–80 who were enrolled in the Group Health Cooperative of Puget Sound sought to determine these women's reasons for initiating, discontinuing, or refusing HRT. The reasons cited most frequently by current users were menopausal symptoms (47.3%), osteoporosis prevention (32.4%), and a physician's advice (30.3%). The rationales offered for quitting HRT included side effects, a physician's advice, fear of cancer, and avoidance of potential menstrual symptoms or bleeding. Of past users, 53.8 percent reported stopping HRT on their own, and 46.2 percent on the advice of their physicians. The most commonly cited reasons for never using HRT were that such hormones were not needed (49.9%), and that menopause was a natural event (17.9%). To the chagrin of the investigators, many women made decisions about HRT independent of their interactions with health care providers.[16]

Those who studied females' reluctance to follow preventive interventions against osteoporosis noted that most women were disinterested in the disease, unless they had salient experiential knowledge of it. Women's lack of interest in osteoporosis possibly represented their opposition to a further medicalization of their lives. The implication of most studies was that women would benefit from better counseling with health care providers. Such views were common, reflecting the widely held beliefs that the management of osteoporosis was the responsibility of medical professionals and that women lacked the knowledge to make informed choices on their own.[17]

BISPHOSPHONATES

In the mid-1990s, the primacy of ERT was challenged by the introduction of bisphosphonates. In the 1960s, during a postdoctoral year at the University of Rochester, Swiss physician Herbert Fleisch had begun working on inhibitors of mineralization that were detectable in urine. Initially he studied polyphosphates, and soon came to the realization that pyrophosphate inhibited tissue calcification and bone resorption. Pyrophosphate, however, was prone to enzymatic breakdown. This work led Fleisch to bisphosphonates. In his seminal papers in *Science* and *Nature* in 1969, Fleisch, through both in vitro and in animal experimentation, demonstrated that bisphosphonates were powerful inhibitors of bone resorption and tissue calcification. Although never involved in clinical trials, he pointed the way to the first use of bisphosphonates, to treat Paget's disease. During the 1970s and 1980s, drug companies evinced little interest in osteoporosis, and Fleisch was disappointed by the lack of progress in applying these compounds as drugs to help alleviate that condition. As late as 1987 he noted that there were "only a few studies" of bisphosphonates as a treatment for osteoporosis.[18] Yet Fleisch's interest in these drugs never lagged. Between 1993 and 2000, he published four editions of *Bisphosphonates in Bone Disease: From the Laboratory to the Patient*, a book that presented data on bone physiology and bisphosphonates.[19]

The lag in evaluating bisphosphonates' new chemistry was partly due to Premarin's popularity, and partly to the fact that osteoporosis was just coming into its own as a disease that affected millions of people both in the United States and abroad. Given Premarin's dominance and the growing size of the market for drugs to treat osteoporosis, it was not surprising that pharmaceutical companies would turn their attention to the development of new ones. In 1994 sales of estrogen and other osteoporosis drugs totaled about $1 billion, and they were expected to reach $3.3 billion by 2000. The European market was even larger, and sales were projected to exceed $6 billion by 1997.[20]

In the 1970s, early results with first-generation bisphosphonate therapy in osteoporotic syndromes proved to be disappointing. For example, one of the drugs in this class, etidronate disodium, had somewhat mixed results. A 1976 study by Robert P. Heaney and Paul D. Saville demonstrated that this drug reduced bone resorption by about 50 percent, yet it also depressed bone mineralization by almost as much.[21] A decade later, several studies employing intermittent cyclical therapy reported much better results. In a prospective two-year, double-blind, placebo-controlled, multicenter study of 429 women, the researchers found that intermittent cyclical therapy with etidronate increased spinal bone mass and reduced the incidence of vertebral fractures in postmenopausal women.[22] Another study, published at the same time, came to similar conclusions.[23] A subsequent follow-up, however, revealed an increase in fracture incidence in the third year. Aside from concerns that etidronate could cause potential demineralization in younger postmenopausal osteoporotic patients when exact dosage regimens were ignored, its use in elderly populations, who required increased amounts vitamin D, could result in occult (hidden) osteomalacia and its accompanying "weak-muscle bone ache" syndrome. Hence the drug was considered investigational at best and did not receive FDA approval as a therapy for osteoporosis.[24]

In the late 1980s several other bisphosphonates were under investigation as possible therapies. By this time, pharmaceutical firms had become aware of the mass market potential for these drugs, and the companies were aggressively supporting research on alternatives to estrogen. Their interest in osteoporosis reflected two distinct objectives. First, the manufacturers hoped that new drugs would reduce the burden that fractures posed to society. Second, they knew that osteoporosis was a chronic (long-term) disease. To prevent or manage this disease meant that any drug receiving FDA approval would have to be taken for years. Acute infectious diseases, on the other hand, were treated with antibiotics administered for very short periods of time. It was already becoming clear to pharmaceutical firms that the development of new antibiotics had a low priority, because their potential return on investment was limited. Instead, these companies had a compelling economic incentive to develop antiosteoporotic drugs, which would clearly maximize profitability.

A basic hypothesis for bisphosphonate treatment was that the decrease in bone remodeling known to accompany its use would involve a greater reduction in bone resorption than in bone formation. This action would be persistent and would have a beneficial effect, varying according to dosage, the duration of its administration, and the type of bisphosphonate used. The principal value of bisphosphonates would be in prevention, but such drugs might also prove to be effective in the treatment of osteoporosis. Orally administered, the bisphosphonates appeared to be generally well tolerated, although side effects included gastrointestinal upset and diarrhea. "The bisphosphonates' oral route of administration, their relative inexpensiveness, and potentially their apparent safety (as confirmed by current studies)," noted one authority, "are definite assets for a potential prophylactic agent. Undoubtedly the next few years will see increasing utilization of the bisphosphonates in the management of osteoporosis."[25]

By the early 1990s, the Merck pharmaceutical company was well on the road to securing FDA approval for alendronate (later marketed as Fosamax). Several studies demonstrated that the oral administration of the drug was well tolerated and effective. The results led the authors of these studies to call for longer-term treatments with larger clinical populations, in order to fully ascertain the potential efficacy and safety of chronic alendronate therapy.[26]

In 1993 the Fourth International Symposium and Consensus Development Conference met in Copenhagen. One of the major sessions—"A New Therapeutic Intervention for the Management of Osteoporosis"—focused on alendronate sodium. Fleisch opened the session by observing that it was "truly exciting" to see the results of the clinical evaluation of the drug as a "promising new therapy" for the treatment of osteoporosis. He then provided a general overview of the new bisphosphonates.[27] At an EFFO conference a year earlier, he had conceded that many questions remained unanswered. Little was known about the best mode of administration for this class of drugs. Should treatment be discontinuous or continuous? What was the best regimen? Were there differences among the various bisphosphonates in this regard? Nevertheless, in Fleisch's opinion the bisphosphonate class had "an important place in the treatment of osteoporosis."[28]

At the 1993 conference, several participants presented findings that confirmed Fleisch's optimism. Gideon Rodan and his Merck associates' preclinical studies attested to the general tolerability of alendronate and showed that the drug inhibited bone resorption.[29] Charles H. Chestnut III and Steven T. Harris reported on the short-term effect of alendronate on bone mass and bone remodeling in postmenopausal women. Their double-masked controlled study of 65 postmenopausal women (mean age = 51.6 years) found that a short-term (six weeks) oral alendronate treatment was well tolerated and effective in decreasing biochemical markers of bone turnover in early postmenopausal women and in stabilizing spinal BMD over a nine-month period.[30]

The Merck pharmaceutical firm was obviously investing considerable resources in alendronate before the FDA approved the drug. The company sponsored the creation of the Fracture Intervention Trial (FIT) in order to determine the effectiveness of alendronate in reducing the risk of fractures in postmenopausal women. By spring 1993, over 6,000 women had been enrolled in randomized studies at eleven clinical centers. The participants were assigned to two substudies. The first (the Vertebral Deformity Study) involved over 2,000 women who had at least one vertebral deformity, and it was designed to test the hypothesis that alendronate reduced new vertebral deformities during three years of follow-up. The second (the Clinical Fracture Study) included nearly 4,500 women without vertebral deformities. Its goal was to determine whether the drug reduced the rate of clinically recognized fractures of all types over an average of 4.25 years of follow-up.[31] The pharmaceutical firm could hardly have been more pleased than by the remarks of D. C. Anderson in summing up the results of the 1993 International Symposium and Consensus Development Conference. Anderson reported that it seemed "quite extraordinary how well the drug works and how safe it is. . . . We are clearly witnessing the development and application of an exciting new drug to a disease which is on the increase world-wide."[32]

Three years later the FIT Research Group published some of their results. Slightly over 2,000 women (age 56–81) with low femoral-neck BMD and at least one vertebral fracture were randomly assigned to two equal groups. One group was given a placebo, and the other, alendronate. New vertebral fractures were the clinical endpoint. After three years, 78 (8%) of the women in the alendronate group had one or more morphometric vertebral fractures (changes in the shape of the bone), compared with 145 (15%) in the placebo group. The study conceded that most vertebral fractures were asymptomatic, as only a third of the radiographically defined vertebral fractures could be recognized clinically. Twenty-two (2.2%) of the women on a placebo had a hip fracture, compared with 11 (1.1%) on alendronate. The investigators admitted that the study had several limitations: most (97%) of the participants were Caucasian, thus omitting other significant groups; it could not address the question of the effects of long-term treatment; it could not determine how long treatment should be continued; and it could not specify what happened to the risk of fractures after discontinuation of the treatment. Nor were the results applicable to women living in institutions or those in poor health. Nevertheless, it seemed clear that alendronate substantially reduced the risk of fractures, including hip fractures.[33]

It was hardly surprising that there was dissent over the findings of the FIT group. Shortly after the article appeared, Adrian Phillips (a British public-health professional) published a critical letter in *Lancet*. He noted that the trial showed a 50 percent reduction in the relative risk of hip fractures (the most important clinical endpoint), but only a 1.1 percent reduction in absolute risk. This meant that 300 patients would require treatment for one year to prevent just one hip fracture. Thus the cost of avoiding that one hip fracture would be about $162,000, a considerable sum. Moreover, few people with trabecular bone fractures were diagnosed as having osteoporosis, thus raising questions about the entry criteria for the study. Nor was there widespread agreement on the value of bone densitometry. "There is still considerable uncertainty about the relation between particular values and risk, as well as the reproducibility of results and quality control."[34]

Phillips's critique did not go unanswered by the investigators. They insisted that his estimate of the cost of reducing one hip fracture was incomplete, since the use of alendronate also prevented sixteen vertebral fractures in seven women and wrist fractures in two others. Furthermore, the predictive value of bone density was at least as good as other measurements that assessed lipid levels (cholesterol) or blood pressure. Hence the FIT results demonstrated that among older women with established osteoporosis, alendronate treatments could reduce the incidence of fractures, including the devastating consequences of hip fracture.[35]

In the FIT group's second study on the effects of alendronate on the risk of fractures in women with low bone density but without vertebral fractures, the investigators found that the drug increased BMD at all sites and reduced the number of clinical fractures from 312 in the placebo group to 272 in the intervention group. Twenty-four women in the placebo group had hip fractures, compared with nineteen in the alendronate group. Although alendronate increased BMD to a similar degree, regardless of the bone's initial density, during the study period there was no significant decrease in the risk of clinical fractures in non-osteporotic women. Still, questions remained. How long should alendronate be administered? Its effects after four years deserved examination, if only because its antifracture efficacy among women without osteoporosis was unknown. It was clear that alendronate accumulated in bone and recirculated when bone containing the drug was remodeled. Thus benefits might continue even after cessation of the therapy. But if the drug caused an adverse effect, endogenous exposure would also continue after the treatment was stopped. Nevertheless, the two FIT studies (both of which were funded by Merck) clearly supported the efficacy of alendronate.[36]

The FIT studies and other investigations, however, also raised serious questions. In a commentary on the FIT studies, Heaney was by no means supportive. He noted that bone measurement technology had proliferated, and physicians were faced with healthy individuals who brought them printouts showing that they had low bone mass. Under normal circumstances the physician treated the patient, rather than the test. When it came to prevention, however, the physician had only the test result. The problem, Heaney stated, was that BMD could not predict fragility fractures. Making a diagnosis solely on the basis of bone mass was inadequate. After adjusting for BMD, other factors—age, a history of fractures after age 40, and a maternal history of hip fractures—were better predictors of fractures than low bone mass. A positive change in bone mass should mean some reduction in the risk of fractures. But that relationship was not necessarily evident, since an increase in BMD did not reverse the loss in strength caused by severed trabecular struts. From a clinical standpoint, Heaney claimed, the most important issue was whether bisphosphonates reduced fragility in individuals with low bone mass who did not exhibit bone fragility. The answer to this question was "maybe." The only certain conclusion was that this class of drugs protected against the development of asymptomatic vertebral deformities and further bone loss. It was unfortunate that osteoporosis researchers had focused exclusively on bone mass, Heaney observed, primarily because it could be measured and because clinicians could then affect it. Fragility, however, involved more than a simple reduction in bone loss. Hence clinicians had no way of distinguishing which patients with the same bone mass had greater or less fragility. "For now," Heaney concluded, "with the exception of asymptomatic spine deformities, the antifracture benefit of bisphosphonates in women with low bone mass but without prevalent fracture must be judged to be small." Heaney, however, was not a therapeutic nihilist, and he pointed out that selective estrogen receptor modulators, HRT, and calcium and vitamin D supplements could also protect against bone loss and prevent fracture.[37]

Equally significant, the statistical results of the FIT trial studying the effects of alendronate on fracture risks in women with low bone density but without vertebral fractures were not as impressive as the Merck pharmaceutical company claimed. The trial enrolled 4,432 women who had bone density measurements 1.6 SD below the mean of young, adult white women. The occurrence of vertebral fractures visible on x-rays was 3.8 percent in the placebo group and 2.1 percent in the treatment group. This equated to a 44 percent relative reduction in the risk of such fractures. In terms of absolute risk reduction, however, the figure was only 1.7 percent—a much more modest decrease. Relative risk—the figure often used by pharmaceutical firms in publicizing the dramatic efficacy of a therapeutic intervention—is misleading. Impressive reductions in relative risk often conceal much smaller reductions in absolute risk. Researchers and drug companies understandably preferred to report results in terms of relative risk. From an ethical and moral standpoint, however, the use of relative risk figures was questionable, if only because it created unrealistic expectations about the effectiveness of a drug.[38]

The FIT studies were confined to the United States, but Merck was also interested in marketing alendronate throughout the world. To this end it created and funded the Fosamax International Trial Study Group (FOSIT), which enrolled 1,908 osteoporotic postmenopausal women from thirty-four countries. Each participant orally took either 10 mg of alendronate or a placebo

once daily (in the morning) for one year. The results were similar
to those of the FIT studies. Alendronate was well tolerated and
produced significant, progressive increases in BMD at the lumbar
spine and proximal femur.[39] A second FOSIT study found that
the drug increased trabecular and total bone density at the ultra-
distal radius (a clinically relevant fracture site).[40] A third study
claimed a 47 percent (relative) risk reduction in non-vertebral
fractures for alendronate-treated patients (nineteen fractures in
the treated group vs. thirty-seven in the placebo group).[41] The
results of these studies strengthened Merck's marketing interna-
tionally.

In a 1996 review of the clinical evidence for alendronate treatment, Englishman Peter Selby (a Fellow of the Royal College of
Physicians, and highly regarded in the osteoporosis community
of researchers) was enthusiastic. Selby noted that alendronate preserved bone mass at all of the sites of major osteoporotic fractures;
it was not associated with any abnormality in bone mineralization; it reduced vertebral and other fractures; and it was well tolerated. While alendronate's long-term safety had yet to be determined, it appeared to be an appropriate alternative to ERT.[42]

On September 29, 1995, the Merck pharmaceutical company
received approval from the FDA to market alendronate as a treatment for osteoporosis and Paget's disease. This action came at a
most propitious moment for the company, given the many uncertainties about the therapeutic efficacy of ERT. Merck thus had an
opportunity to publicize the virtues of alendronate as an alterative
therapy that would both alleviate many of the problems associated with osteoporosis, yet avoid the risk of adverse side effects.

Even before receiving FDA approval, Merck had begun to
prime the market. It cosponsored a campaign with the NOF, urging women to undergo diagnostic screening. To distinguish the
company's drug from estrogen, Merck kept referring to what it
called its "non-hormonal treatment." The prospect of a block-
buster drug led to a 14 percent increase in the price of the pharmaceutical firm's shares between May 15 and June 14.[43] When the
drug became available, Merck intensified its marketing activities.

"Consumer research," the company noted in its annual report for 1995, "shows a growing level of awareness and knowledge about osteoporosis. We have been aggressively working to educate consumers about the disease and the importance of early diagnosis and appropriate treatment with organizations such as the National Osteoporosis Foundation, the Older Women's League, and members of the European parliament. We have established the Bone Measurement Institute, a nonprofit organization, to increase the accessibility and affordability of bone measurement technologies. Merck has also provided funding for the first World Summit of Osteoporosis Societies. . . . We believe that our marketing and research programs on osteoporosis will ultimately benefit millions of people, significantly reduce disabilities that lead to long-term care, and improve the quality of life." Between 2000 and 2003, sales of Fosamax increased from $1.2 billion to $2.7 billion, and it was the most frequently prescribed drug for osteoporosis.[44] Many of the studies of the therapeutic qualities of alendronate that appeared in medical journals were funded by the Merck pharmaceutical company, which also owned and funded NORA, FIT, and FOSIT. To foster public awareness, the firm ran television commercials that played on the fears of older women to make them receptive to the use of Fosamax.

The entry of alendronate into the marketplace immediately set off a high-stakes battle. American Home Products Corporation (the parent company of what was now Wyeth-Ayerst) launched a powerful campaign to defend Premarin, which accounted for three-quarters of the estrogen-therapy prescriptions in the United States. Although primarily used to mitigate menopausal symptoms, Premarin had also been marketed as an antiosteoporotic drug. Faced by a new and potentially powerful competitor, American Home Products claimed that most studies found no link between breast-cancer risks and estrogen. "The medical community," the company told a *Wall Street Journal* reporter, "has long recognized the long-term benefits of [estrogen replacement therapy], including the prevention of osteoporosis, a reduction in the incidence of associated hip and wrist fractures by 60%, and the potential reduction of coronary heart disease by 50%." It also lined up prestigious scientists to rebut the study by Harvard researchers, published in the *New England Journal of Medicine* in 1995, claiming that ERT increased the risk of breast cancer. Dr. Leon Speroff (of the Oregon Health Sciences University), who had done clinical-trial work for the Merck pharmaceutical company, urged caution in interpreting the Harvard study. "As a physician/researcher who treats and studies menopausal women," he wrote, "I've witnessed first hand the hysteria that can be generated as a result of unbalanced study data—particularly data related to breast cancer." American Home Products also disseminated an NIH statement about the Harvard study. The NIH noted that there had been more than thirty studies examining a possible link between ERT and breast cancer, with conflicting results. NIH's statement noted that there was "very little or no overall risk" of breast cancer with the use of ERT (or HRT), although some studies suggested a link with prolonged use. Ongoing clinical trials might provide clearer answers.[45]

The battle between Merck and American Home Products was a portent of future developments. By the mid-1990s, more than twenty antiosteoporotic drugs were in varying stages of development, about half of which were non-estrogen compounds. Among them was Eli Lilly and Company's raloxifene, which received FDA approval in 1999 for the treatment of osteoporosis. Eight years later the FDA approved a new use of this drug: to reduce invasive breast-cancer risks in postmenopausal women with osteoporosis and postmenopausal women at high risk for breast cancer. After 2000, a variety of other antiosteoporotic drugs entered the medical armamentarium. Given the potential size of the market, pharmaceutical firms spent millions to draw attention to the disease, to urge women to undergo screening for osteoporosis, and above all to use their drugs.

MARKETING DRUGS AND SCREENING WOMEN

As new drugs were developed, screening assumed an ever more central role. Screening would identify more asymptomatic women, which in turn would expand the market for pharmaceuticals and imaging equipment. It would also create new opportunities for physicians in a variety of specialties, including gynecology, orthopedics, radiology, endocrinology, and family and internal medicine. Given that reimbursement systems rewarded tests and interventions, there was little doubt that many clinicians would seize an opportunity that presumably aided their patients while enhancing their earning potential.

By the mid-1990s, the National Osteoporosis Foundation was playing an increasingly prominent role in warning about the health threats posed by osteoporosis and the need for screening and treatment for this disease. With financial support from the Merck pharmaceutical company, it convened a panel of expert clinicians to assess the contributions of osteoporosis to particular types of fractures among different populations in the United States. Given the paucity of specific data, it employed the Delphi technique to estimate the probability that each of the seventy-two categories—consisting of four fracture types (hip, spine, forearm, and all other sites combined), three age groups, three racial groups, and both genders—was associated with osteoporosis. The Delphi technique (originally developed by the Rand Corporation in the 1950s) was a method for achieving convergence of opinion concerning real-world knowledge solicited from experts within certain topic areas. This technique, which was used widely in the medical arena, was based on the assumption that a systematic, literature-based, scientific method that utilized group judgment in the absence of adequate empirical data would nevertheless result in reliable knowledge. A point that was never considered was that critics of osteoporosis as a disease concept were excluded from this expert panel.

The consensus of the panel was that previous researchers had underestimated costs. Panel members assessed health care expenditures attributed to osteoporotic fractures in 1995 at $13.8 billion, of which $10.2 billion (75.1%) was for the treatment of white women, $2.4 billion (18.4%) for white men, $0.7 billion (5.3%) for non-white women, and $0.2 billion (1.3%) for non-white men.

Of this total, 62.4 percent was for inpatient care, 28.2 percent for nursing home care, and 9.4 percent for outpatient services. The panel warned that the aging of the American population could lead to large increases in future expenditures.[46]

A second panel, also employing the Delphi technique, was convened by the NOF to make judgments about the probabilities that fractures of different types might be related to osteoporosis, categorizing them according to the patients' age, gender, and race. This panel's estimates were also higher than previous ones. On a scale ranging from 0.00 (no attribution) to 1.00 (100% attribution), the median attribution probability for hip fractures for women age 45–64 was 0.75; for those 65–84, 0.85; and for those older than 85, 0.95. Panel members conceded that it was not possible to rigorously validate any of their osteoporosis attribution probabilities, but they insisted that such figures were needed to provide the basis for a cost-effectiveness analysis that could be used to develop a health care policy perspective on interventions to prevent osteoporotic fractures.[47]

A year later the NOF, in collaboration with nine other organizations (including the American College of Obstetricians and Gynecologists, the American Geriatrics Society, the American Academy of Physical Medicine and Rehabilitation, and the American Society for Bone and Mineral Research) published the first edition of its *Physician's Guide to Prevention and Treatment of Osteoporosis*. Its recommendations were startling, largely because they dramatically expanded the target population for medical and pharmaceutical intervention. All women, the NOF insisted, required counseling about risk factors for osteoporosis. All postmenopausal women should be evaluated using BMD testing to confirm the diagnosis and determine disease severity. Such testing should include postmenopausal women under the age of 65 who had one or more risk factors, and all women over the age of 65 regardless of additional risk factors. Patients were urged to take at least 1,200 mg per day of dietary calcium and 400–800 IU per day of vitamin D. It endorsed regular weight-bearing and muscle-strengthening exercises and advised patients to avoid smoking and to keep their alcohol intake at a moderate level. All postmenopausal women with vertebral or hip fractures should be considered candidates for treatment. Therapy to reduce fracture risk was indicated for women with T-scores below −2 in the absence of risk factors, and those with T-scores below −1.5 if other risk factors were present. Pharmacological options for prevention and/or treatment included HRT, alendronate, and raloxifene, with calcitonin used only for treatment.[48]

The NOF guidelines reflected an aggressive approach that was not necessarily shared in other nations. The strategy of the NOF was to identify all individuals who could receive treatment for $30,000 / quality-of-life years saved or less. Its ubiquitous guidelines, noted three well-known British physicians, targeted a huge population for assessment and treatment. The NOF guidelines—in contrast to the much more conservative ones of the EFFO—bordered "on the evangelical and are difficult to justify from the principles articulated in their own resource document."[49] A position statement by the IOF noted that the vast majority of fractures occurred among individuals deemed at low risk, indicating that the screening test for BMD, used just by itself, had low sensitivity.[50]

There were also significant national differences in perceptions about and treatments of osteoporosis. In Europe, for example, ERT was rarely prescribed for osteoporosis. The use of calcium varied widely, ranging from 80 percent in Germany to 25–28 percent in Italy. One-third of the prescriptions given for osteoporosis in the United Kingdom were simple analgesics, perhaps because British physicians did not believe in the efficacy of drugs, or else because they believed that osteoporosis could be treated solely with an analgesic. "There are no current clinical guidelines for the diagnosis and treatment of osteoporosis in Europe," noted one authority, "only areas of common perception."[51] In Japan, only elderly women with established osteoporosis were treated, largely because women in the immediate postmenopausal period were virtually free of the symptoms (such as lumbago) and signs (such as spinal compression fractures) of this condition. The Japanese emphasis on the use of calcium, vitamin D, and calcitonin reflected a "fatalistic and naturalistic view toward menopause; [and] an uneasiness with hormone replacement therapy."[52]

Therapeutic guidelines were far more aggressive in the United States than elsewhere. The NOF was obviously a driving force in suggesting a radical expansion of both screening for and the pharmacological treatment of osteoporosis. To manufacturers of BMD testing equipment and the pharmaceutical industry, such recommendations were greeted with unbridled enthusiasm. If implemented, they would have a dramatic effect on product use. Thus it was hardly surprising that pharmaceutical firms would be extraordinarily generous in their financial support of the NOF,[53] researchers, clinicians, epidemiologists, and those conducting clinical trials. The profit potential was huge, given the fact that osteoporosis drugs would have to be taken daily for many years, at costs ranging from $400 to $1,000 per annum.

U.S. reception of the NOF guidelines was generally positive. Bernadine Healy, a former director of the NIH, wrote an enthusiastic editorial in the *Journal of Women's Health*. Osteoporosis, she insisted, "is not an inevitable part of aging." Knowledge about risk factors—the most important of which were age, the female gender, and menopause—was the first step in understanding this illness. Women's bones deteriorated when deprived of estrogen and progesterone, Healy averred. Since women were born with a thinner-framed, lower-weight body type, their risk was multiplied. BMD testing, which had moved from its prior major use as a research tool to become an indispensable agent in patient care, constituted a major breakthrough. In addition, she noted, options for the prevention and treatment of osteoporosis had multiplied. As more and more women lived into their ninth and tenth decades, the consequences of osteoporosis would become more visible. "Women have achieved equality because of brain power—not because of muscle power or brawn," Healy concluded. "It is thus sadly ironic that in the end women face a large risk of being overtaken by a loss of bone and brawn. The message of the National Osteoporosis Foundation is that those weak and feeble bones need be no more."[54]

Donna F. Heinemann (of the State University of New York at Stony Brook) was also positive in her review of the NOF guidelines. In the past, she noted, many studies used a change in BMD as surrogate data for fractures. More recent prospective, controlled clinical trials that used fractures as an endpoint yielded valuable data. Yet this approach had some limitations, observed Heinemann. New vertebral fractures were difficult to diagnose; there were many clinical definitions and interpretations of what constituted a vertebral fracture; and many vertebral fractures were asymptomatic. Because hip fractures were symptomatic, they were easy to track. She stated that future studies would probably use both fractures and a change in BMD as endpoints. Nevertheless, Heinemann noted that the NOF recommendations had limitations. Data for groups other than Caucasian women were lacking. Many of the clinical trials were only concerned with medical outcomes (e.g., a decrease in the incidence of fractures and changes in BMD), and failed to take into account ancillary endpoints, such as pain, a limitation in the activities of daily living, involvement of the secondary organ system, or psychological disturbances (e.g., depression). These limitations notwithstanding, Heinemann believed that the NOF guidelines offered the most comprehensive approach to the prevention and management of osteoporosis.[55]

The NOF recommendations proved to be influential, and they played an important role in hastening the expansion of screenings for osteoporosis and prompting the consequent increase in the use of drug therapy for this condition. Yet faith in bone density screenings was not universally pervasive. Around the same time the NOF guide was published, the Swedish Council on Technology Assessment in Health Care and the British Columbia Office of Technology Assessment issued reports that differed in fundamental respects from the NOF guide.

The Swedish report, entitled "Bone Density Measurement—a Systematic Review," appeared in 1997 as a supplement to the highly regarded *Journal of Internal Medicine*. Prepared by a coalition of well-known Swedish investigators and reviewed by an equally prominent group, the report raised serious questions. The key issues concerned the reliability of the various methods to measure bone density and whether bone density treatments could actually prevent fractures. In a meta-analysis and a review of the literature, the group conceded that a reduced bone density level might be defined as osteoporosis, but it was uncertain if a lower level actually promoted fracture. "Bone density measurement does not provide reliable evidence that a fracture will occur at some time in the future, because low bone density values are only one of several risk factors. Most fractures result from accidents, often falls." The report stated that some individuals with normal values would still experience fractures, while others with low values would not. Nor was there a clear threshold between normal and morbid values. Contemporary technologies demonstrated good precision and used a small dose of radiation, but their accuracy was low (approximately 10%). Nor was there much experience with treatment results, particularly when osteoporosis was combined with other diseases.

The conclusion of the Swedish investigators was clear and unequivocal. "The scientific evidence is insufficient to recommend bone density measurement in mass screening of asymptomatic individuals, including mass screening of women following menopause and opportunistic screening of patients who present no symptoms of osteoporosis but who contact health services for other reasons." The report noted, however, that there were a few situations in which bone density measurement was warranted: for patients with primary diseases that increased the risk of fractures; for patients with vertebral fractures; for patients with diseases treated with medications that increased risks to the skeleton (e.g., cortisone therapy); as a way to monitor the results of treatment, provided that this was done under controlled conditions and with tests repeated only at intervals greater than two years; and in research projects.[56]

Two British Columbia Office of Technology Assessment reports posed even broader challenges to the prevailing belief that osteoporosis was a disease that required BMD screening for its management and prevention. The first report (in December 1997) dealt with a major issue: did the evidence support the selective use of BMD testing in well women? After an exhaustive analysis, the report emphasized that BMD measurements were poor predictors of future fractures, and that their continued use would misdirect treatment efforts away from the majority of women who would ultimately suffer fractures by focusing attention on the minority with low bone density. If women with a T-score of –1 were classified as being at increased risk for fractures, then 70 percent of the women who would ultimately suffer fractures would not come from the group identified as being at risk. The group at greatest risk for life-threatening and debilitating hip fractures were women of advanced age (80–89) with multiple risk factors, regardless of their BMD.[57]

The 1997 report also pointed out that the assumption that the various BMD testing technologies were accurate and replicable was not justified. There was no industry-wide agreement on one specific standard that would provide machine-to-machine comparability. Nor was the assumption that the current cohort of women approaching menopause would have the same burden of suffering due to fragility fractures as the cohort of women currently in the age groups with high incidences of fractures. The bones of women born in the first two decades of the twentieth century reached their peak in the early 1940s—a period that spanned two world wars and the Great Depression. Patterns of diet and exercise differed in significant ways in that generation. The current cohort of women entering menopause could have significantly different fracture rates, compared with their mothers' generation.[58]

The authors of the first British Columbia report also emphasized the importance of studying clinical medicine within a social and political context. Medicine did not merely describe a preexisting biological reality; it also reflected societal beliefs, values, and relations of power. In American and western European societies, both aging and menopause had become increasingly medicalized; a natural phenomenon had been transformed into a disease state, thus heightening the public's anxiety. Under these circumstances, the pharmaceutical industry promoted the view that estrogen therapy was vital to the health of women age 40 and over, to say nothing of its financial support to consultants who advised national governments on policy issues, and forums that redefined osteoporosis using BMD measurement thresholds. Market forces also created and capitalized "on a climate of risk and reassurance," which then drove the deployment of health technologies regardless of whether they improved health outcomes. In summarizing the social and political contexts of medical care, the report emphasized that when cultural and social expectations intersected with new technologies, "a strong force to 'normalize' interventions is created; these then diffuse into clinical practice before evidence is available about their actual usefulness. BMD testing is in effect being promoted and accepted as a prophylactic to counteract the primarily social attitudes towards aging (manifested in images of decaying bodies, loss of mobility, lifelong dependency), at the expense of examining the political and economic implications of aging on women's health."[59]

The 1997 report concluded by insisting that selective use of BMD testing in well women was not supported by any scientific evidence. Clinical-trial evidence did not link BMD tests with improved outcomes. BMD screening had only a limited ability to identify women who would suffer fractures in the future; its contribution to overall risk assessment had not been established; and some evidence suggested that BMD screening could create a harmful response in the women being tested. Clinical practice, moreover, had not been altered by BMD measurements. Finally, models used to estimate the impact of BMD testing on a population's health found only a slight effect. Since the long-term benefits of screening remained speculative, the report stated that "significant harm may exist with the use of BMD as a screening strategy."[60]

In a much briefer report a year later, the British Columbia Office of Health Technology Assessment insisted that passage through the life cycle was both a social and a biological process.

Yet the medical focus was increasingly confined to such biological processes as "failing ovaries." Natural phenomena became labeled as diseases, thus leading to the marketing of fear. BMD testing, the report continued, was not the only technology driven by a market-dominated climate that emphasized risk and reassurance, regardless of whether such technologies led to improved health outcomes: prenatal ultrasound, electronic fetal monitors, predictive genetic and other screenings, and hormone therapy were comparable examples. "Being 'at risk' has come to mean being 'diseased.' On-going medical intervention has accordingly come to be seen as necessary to prevent the body from aging."[61]

The medical view of menstruation as a pathological condition and the disagreeable events associated with female aging—both of which were intimately related to the diagnostic category of osteoporosis—also came under fire from critics who insisted that such interpretations were based on misconceptions and claims that bore no relationship to reality. One of their arguments was that the medical view of aging reflected a belief that in the past women—like most mammals—did not live beyond the age of menopause. Technological and cultural forces, however, favorably influenced longevity. The result was a claim that women who lived past reproductive age were "biological anomalies." Based on an ignorance of demographic data, such claims, buttressed by pharmaceutical advertising, led to the belief that menopause was a condition in need of medication (HRT). In a comparative study of the United States, Japan, and Canada, Margaret Lock and Patricia Kaufert (two figures well-known for their writings about menopause) challenged the notion of a universal menopause. They found that both the symptoms reported at menopause and postmenopausal disease profiles differed from one population to another. Similarly, mortality and morbidity data from these societies demonstrated that postmenopausal women were not equally at risk for heart disease, breast cancer, or osteoporosis. The end of menstruation, Lock and Kaufert emphasized, should not be conceptualized as an invariant biological transformation. It was more appropriate to think of " 'local biologies,' which reflect the very different social and physical conditions of women's lives from one society to another." Their findings, these authors concluded, clearly had profound implications for the medical claim that menopause was a pathological condition that required a lifetime course of HRT.[62]

The osteoporosis community, the pharmaceutical industry, and manufacturers of imaging equipment remained all but oblivious to such criticisms. Although disputes among clinicians and investigators were common, there was no disposition to challenge the definition or legitimacy of the diagnosis of osteoporosis. In the first edition of the book *Osteoporosis* (edited by three well-known figures and soon to become a standard work on the subject), the preface commented in approving terms on the recent explosion of interest in and information about osteoporosis. It noted that the traditional view, first promulgated by Fuller Albright in 1948, was that osteoporosis was a specific disease entity, resulting from an exaggerated loss of bone following menopause or changes associated with progressive aging. A more recent formulation was that osteoporosis simply meant a condition of global skeletal fragility, at a level sufficient to increase the risk of fractures. Thus a person's skeletal condition reflected a variety of genetic and environmental influences. This paradigm made it easier to accept the WHO recommendation that osteoporosis could be considered to be an established medical condition purely on the basis of a bone mass measurement. Each chapter in the book—prepared by an expert—described the lifelong skeletal effects of genetic and environmental influences and provided even skeptical readers with a solid basis from which they could reach their own decisions. The volume ran to over 1,300 pages and included seventy chapters. Five years later a larger, two-volume second edition, with seventy-nine chapters, was published.[63]

Virtually all of the chapters in the 1996 edition (as well as in the 2001 second edition) assumed the legitimacy of the diagnosis. For example, in their discussion of the magnitude and scope of osteoporosis and fractures, Cyrus Cooper and L. Joseph Melton III emphasized that osteoporosis was "a complex, multifactorial chronic disorder in which a variety of pathophysiologic mechanisms lead to a progressive reduction in bone strength, and an increased risk of fracture." Only in recent years, they stated, had the precise burden of this condition been recognized. At least one third of postmenopausal white women in the United States had an extremely low bone mass, with their lifetime risk of fracture from age 50 onward approaching 40 percent. Moreover, these authors noted, the future burden and costs of osteoporosis could only rise in future generations as the number of older people in the population increased. Such demographic changes would cause the number of hip fractures occurring among people age 35 and over throughout the world to increase from 1.66 million in 1990 to 6.26 million in 2050. In the opinion of both authors, measures were urgently required to avert this trend.[64]

That same year John A. Kanis (a highly regarded British medical professor, researcher, and WHO advisor) published his *Textbook of Osteoporosis*. He also ignored those who were critical of the dominant paradigm. By 1996, when his textbook appeared, much was known about the pathophysiology and treatment of osteoporosis. There were many regimens that prevented the progression of this disease, but they did not offer a cure. The major problem was that a diagnosis of osteoporosis was made only after a fracture had occurred. Nevertheless, Kanis believed the disease was largely preventable, and he focused on two strategies. The first involved the identification of women with low BMD, women likely to fall, and women with certain diseases. The second was a population-based strategy, intended to modify a risk factor within the general community. Kanis pointed out that if BMD was increased by 10 percent among women, fragility fractures in females would be reduced by 50 percent. Several risk factors—if causally related and correctable—might have a significant impact. Thus higher levels of exercise, smoking cessation, high-calcium diets, and the universal use of HRT by postmenopausal women would reduce risk. Nevertheless, he stated, not all of these factors were necessarily causally related. Nor was it clear that maladaptive behaviors could be modified. The universal use of HRT could reduce the burden of fractures in postmenopausal women by 50 percent or more. But most of these women did not take HRT, and nothing was said about what the risk and benefit calculus would be if they did so. Kanis noted that it was also uncertain whether the effects of HRT persisted, and whether catch-up bone loss occurred. HRT was recommended for ten years, but the vast majority of fractures occurred after age 70. His discussion indicated the problematic character of most interventions: prevention was a laudable goal, but the means were beset with difficulties that were not easily resolved.[65]

As the twentieth century came to a close, it was clear that osteoporosis had risen to an important position in the pantheon of diseases. The early twentieth-century interest in bone physiology and endocrinology that led to Albright's pioneering work in the 1940s underwent a fundamental transformation during and after the 1980s. The creation of an international community of osteoporosis researchers and clinicians, the growing influence of the pharmaceutical industry and manufacturers of imaging machines, the development of new drugs, the creation of national and international organizations, and the activities of federal agencies in convening consensus conferences and supporting research all contributed to the elevation of osteoporosis as a legitimate diagnostic category. Less visible, but equally significant, was the persistence of a long-held belief that menopause, far from being a normal part of the life cycle, instead was a deficiency disease that required medical management. The evidentiary and epidemiological foundations on which the diagnosis of osteoporosis rested left much to be desired, but these were rarely discussed. Moreover, the enthusiastic claims of therapeutic efficacy in treating this condition were often exaggerated. Yet there were few checks on a coalition bent on heightening the importance of osteoporosis. In the first years of the twenty-first century, the preoccupation with this diagnosis would reach new heights.

CHAPTER SEVEN

Osteoporosis Triumphant?

In July 1990, President George H. W. Bush signed a proclamation designating 1990–2000 the Decade of the Brain. The goal was "to enhance public awareness of the benefits to be derived from brain research." The designation resulted in an extraordinary increase in the visibility of neuroscience and bipartisan congressional support for the NIH in general and neuroscience in particular.[1] Hoping to emulate the successes of the brain decade, a group that met in Sweden in 1998 agreed to launch the Bone and Joint Decade 2000–2010. Its aims were fully in line with those of organizations such as the NOF and the IOF. Launched officially in 2000, the project quickly secured the endorsement of the WHO.[2] The United States Bone and Joint Decade 2002–2011, officially proclaimed by President George W. Bush, was endorsed by all fifty states; numerous national health care professional, patient, and public organizations; and every medical school followed suit.

Interest in and concern with osteoporosis was attaining a level hitherto never reached. During the first decade of the twenty-first century, the federal government, the WHO, and national and international organizations labored to increase public awareness of the dangers of osteoporosis; the influence of the pharmaceutical industry reached a new peak as it expanded the market for its drugs, and provided generous financial support for both researchers and clinical trials; manufacturers of imaging equipment prospered as sales grew rapidly; and bone densitometrists and bone technologists increased their influence by developing new diagnostic criteria that made screening an indispensable tool. Warnings about the worldwide "epidemic" of osteoporosis proliferated, and new treatment guidelines began to include virtually the entire population of older citizens.

The publicity accorded to the dangers of osteoporosis and the benefits of treatment, however, concealed disquieting elements. The claims that the aging of the population, both in the United States and elsewhere, would lead to more fractures began to be contradicted by new epidemiological data. Despite the increasing use of pharmacotherapy, follow-up data suggested that many of these medications were neither as safe nor as effective as their supporters claimed. Nor did data necessarily support the claims that screening for osteoporosis was effective in identifying individuals at risk for fractures. Critics also pointed to many of the detrimental consequences that followed the pharmaceutical industry's ability to co-opt both researchers and clinicians and to publicize claims of the alleged benefits of their drugs, even in the absence of persuasive evidence.

HARBINGERS OF CHANGE

The heightened preoccupation with osteoporosis was reflected in the growing concern of the U.S. Congress and the NIH about this disease. In 1999 a Senate subcommittee held a special hearing to discuss osteoporosis and explore the role that the federal government could play in increasing research, education, and preventive efforts. Testimony by federal health officials, representatives of the NOF, and several members of Congress emphasized the importance of the problem and the need for greater research funding.[3] A year later the National Institute of Arthritis and Musculoskeletal and Skin Diseases and the NIH Office of Medical Applications of Research sponsored a three-day consensus conference dealing with osteoporosis prevention, diagnosis, and therapy. Six other NIH institutes also cosponsored the conference, together with NIH's Office of Research on Women's Health and the Agency for Healthcare Research and Quality. A thirteen-member panel was convened, representing the fields of internal medicine, family and community medicine, epidemiology, orthopedic surgery, gerontology, rheumatology, obstetrics and gynecology, preventive medicine, and cell biology. Thirty-two experts from these fields presented data to both the panel and 699 audience members. Prior to the conference, a bibliography of 2,449 articles (derived from the National Library of Medicine's MEDLINE/PubMed database) had been provided to the panel, from which its members developed a draft statement that was then circulated to the experts and the audience at the conference for public discussion. After resolving conflicts, the panel released a revised statement at the end of the conference.[4]

That statement was largely a summation of existing knowledge and a presentation of guidelines for future research. It began by repeating the familiar refrain that osteoporosis was a "major threat" to Americans. At least 10 million people had the disease, while another 18 million had low bone mass, which placed them at increased risk. The conference statement then addressed five key questions. What was osteoporosis and what were its consequences? How did risks vary among different segments of the population? What factors shaped skeletal health throughout the life course? What was the optimal evaluation and treatment of osteoporosis and fractures? What directions should future research take?

The statement defined osteoporosis as a skeletal disorder characterized by compromised bone strength, predisposing a person to an increased risk of fractures. Bone strength reflected the integration of bone density (grams of mineral per area or volume) and bone quality (architecture, turnover, damage accumulation, and mineralization). It noted that unfortunately, there was no accurate measure of overall bone strength. The WHO definition of osteoporosis (a T-score of −2.5) was controversial, largely because it was not clear how to apply this diagnostic criterion to men and children or across ethnic groups. The difficulties of accurate measurements and of standardization between instruments and sites were a further source of controversy. Osteoporosis (and especially hip fractures) had financial, physical, and psychosocial consequences that affected both individual families and the community.

In addition, the statement continued, the prevalence of the disease varied by gender, race, and ethnicity. Women were far more vulnerable than men, while African Americans had the highest BMD. Bone mass attained in early life was perhaps the most important determinant of skeletal health. Hence nutrition and exercise were vital elements in preserving bone health.

According to the conference statement, the goals for the evaluation of patients at risk for osteoporosis were to establish this diagnosis on an assessment of bone mass, to establish fracture risks, and to make decisions on when to institute therapy. T-scores were originally based on an assessment of BMD by DXA scans. Subsequently, T-scores were applied to define diagnostic thresholds at other skeletal sites and used for other technologies, thus failing to produce comparable data between sites and techniques. The value of universal screening had yet to be determined. Hence, until an assessment by randomized clinical trials was conducted, individual decisions regarding screening for osteoporosis had to be informed by preliminary evidence, which indicated that the risk of fracture rose with age and with an increased number of additional risk factors. Effective treatments included calcium and vitamin D supplementation, physical activity, bisphosphonates, and HRT.

The panel's statement concluded by offering an agenda for future research. It was important to develop strategies to maximize peak bone mass; to identify genetic factors; to evaluate the impact of glucocorticoid-induced osteoporosis in adults and children; to undertake prospective studies of gender, age, and ethnically diverse individuals to more accurately identify fracture risks in these populations; to develop quality-of-life tools that would measure the effects of fracture risks and interventions; to study the psychological impacts of fractures on neuropsychiatric disorders that may cause or be the result of osteoporosis; to better evaluate and manage fractures; to determine the most effective method of educating the public and health care professionals about the prevention, diagnosis, and treatment of osteoporosis; and to determine the efficacy and safety of the long-term administration of drug interventions in maintaining BMD and preventing fractures.

In many respects the NIH consensus conference reflected the concerns and issues characteristic of the 1990s. Although there was general agreement that osteoporosis was a major public health threat, the participants broke no new ground in their analysis or their recommendations. Within a short period of time, however, a series of developments would both transform and complicate the prevailing paradigm about the diagnosis and treatment of osteoporosis.

Two years earlier, the results of the Heart and Estrogen/ Progestin Replacement Study (HERS) were published in *JAMA*. A group at the University of California, San Francisco, had decided to test the claim that estrogen reduced cardiovascular disease by launching a randomized, blind, placebo-controlled secondary prevention trial of women with coronary heart disease (CHD). When the National Heart, Lung, and Blood Institute rejected this proposal, because the institute was about to solicit applications for primary prevention studies under the auspices of the Women's Health Initiative, the group secured funding from the Wyeth-Ayerst pharmaceutical company, which agreed to give the investigators complete jurisdiction over the study, the analysis of the data, and the writing of the report. The firm was interested in testing Prempro, its new drug that combined estrogen and progestin for cardioprotection. Wyeth-Ayerst hoped that if evidence showed that this pill reduced the risk of heart disease, that conclusion could be generalized to estrogen alone, thus expanding the market to all postmenopausal women.[5]

The trial enrolled 2,763 women with coronary disease who were younger than 80 years old and postmenopausal women with an intact uterus. The results were shocking. Hormone therapy did not immediately reduce the overall rate of CHD, and it did increase the risk of blood clots. There was a pattern of an early increase in the risk of CHD events for those receiving HRT, but, after several years of hormone therapy, this shifted to a favorable pattern concerning CHD events. The researchers therefore did not recommend starting HRT for secondary prevention of CHD, but they suggested that women already receiving hormone therapy should continue with the treatment.[6]

Despite its conclusions, the study did not produce any immediate changes in HRT use. Many observational studies and meta-analyses had found that ERT decreased the risk of CHD, but such studies had to be viewed with caution. The findings of the HERS study, noted Diana B. Petitti in an accompanying editorial in the same issue of *JAMA*, "are a sobering reminder of the limitations of observational research, the incompleteness of current understanding of the mechanisms of vascular disease, and the dangers of extrapolation." Nevertheless, Petitti continued, since the beneficial effects of estrogen on bone and menopausal symptoms had been established in randomized trials, there was no emergency, since HERS identified no new risks.[7] Another editorial in the *Journal of Women's Health* pointed out that the HERS trial was consistent with evidence showing cardioprotection in longer-term users. What happened during the first eight months of treatment was unknown, but it was urgent to find that out.[8]

The HERS results did not arouse interest in the medical community, nor did that study receive much publicity in the media. Many found it difficult to believe that a single study could overturn a huge amount of data from other observational studies and clinical trials. Skepticism was widespread, and clinical practice remained untouched by the findings of the study.[9]

The same was not true when the Women's Health Initiative clinical trial of HRT was halted in 2002. Between 1999 and 2001, the board monitoring the data found an elevated number of adverse cardiovascular events, but it cautioned that these were preliminary findings. In May 2002, however, the board found that not only was there an elevated risk of adverse cardiovascular events, but of breast cancer as well. The study was then stopped, three years before its scheduled endpoint. "Overall health risks exceeded benefits from use of combined estrogen plus progestin for an average of 5.2-year follow-up among healthy postmenopausal US women," the WHI study concluded. "The risk-benefit profile found in this trial is not consistent with the requirements for a viable intervention for primary prevention of chronic diseases, and the results indicate that the regimen should not be initiated or continued for primary prevention of CHD."[10]

The same issue of *JAMA* included an article dealing with the relationship between HRT and the risk of ovarian cancer. That study involved a 1979–1998 cohort of 44,281 postmenopausal women in the Breast Cancer Detection Demonstration Project. It found that women who used only ERT for ten or more years had a significantly increased risk for ovarian cancer. Those who used only short-term estrogen-progestin replacement therapy were not at increased risk. Nevertheless, the study urged further investigation of both short- and long-term estrogen-progestin therapies.[11]

The cessation of the WHI study received national publicity. Both television and newspapers reported the results, often in spectacular form. The manner in which the results were reported, as Elizabeth S. Watkins noted, raised questions about "the alleged objectivity and expert vocabulary of scientific experimentation, the public nature and lay language of consumer information, and the art and empiricism of medical practice."[12] Depending on the manner in which they were presented, the actual hazards of HRT could be read in very different ways. The women who took 5.2 years of HRT increased their risk of CHD by 29 percent, breast cancer by 26 percent, strokes by 41 percent, and pulmonary embolisms by 213 percent. Colorectal and endometrial cancers and hip fractures, however, showed a decreased risk. Put in terms of absolute excess risks per 10,000 person-years attributable to estrogen-progestin therapy, there were seven more CHD events, eight more strokes, eight more pulmonary embolisms, and eight more invasive breast cancers, while there were six fewer colorectal cancers and five fewer hip fractures.[13]

The meaning and relevance of these data were anything but clear. Some women felt that the benefits of HRT outweighed the risks of cancer in future decades, while others came to diametrically opposite conclusions. At the University of Florida Women's Clinic, 1,076 women (out of a total of 6,468) completed a questionnaire about HRT: 65 percent had taken the drug for symptom relief, whereas only 16 percent used it for osteoporosis. When asked whether their attitude had changed because of the WHI results, 42 percent replied in the affirmative. Many of the respondents dramatically overestimated the risks of HRT, but, despite their increased caution, only 35 percent indicated that they would not recommend HRT to a friend.[14]

The National Women's Health Network, which for years had fought against the claims by pharmaceutical companies about the benefits of estrogen, waxed enthusiastic about the results of the WHI. At about the same time that articles on the WHI study appeared, the National Women's Health Network published its own analysis of HRT and osteoporosis. It noted that bones did break, but that osteoporosis "may—or may not—have something to do with it." It also argued that osteoporosis was not a disease, "but simply one risk factor for bone fracture in the same way that high cholesterol is one risk factor for heart disease." The network's analysis stated that bones broke for multiple reasons, including falls, and many falls were caused by drugs such as antidepressants, antipsychotics, barbiturates, and tranquilizers. Yet women were constantly told that osteoporosis was caused by menopause, and that screenings for this condition and hormone therapy were the appropriate courses of action. The marketing of drugs such as alendronate by pharmaceutical companies exposed "too many women to drug treatment."[15] Clinicians struggled in their efforts to respond to questions raised by their patients in the wake of all these findings. Nor was the WHI study exempt from criticism; some believed its design was flawed. Whatever the opinions were, the sales of Prempro dropped dramatically as more studies appeared in the months following the publication of the WHI results.[16]

During the controversy over the WHI study, little attention was paid to its methodological underpinning. In scientific work, the claim that a particular pathogen caused a specific infectious disease required a rigorous step-by-step process (generally following the criteria that were first enumerated by Robert Koch and are still known as Koch's postulates) that had to lead to a persuasive demonstration that brooked not even a single exception. Epidemiological studies, on the other hand, gave rise to hypotheses supported by statistical associations (correlations) derived from large samples. There was no chain of linked evidence; contradictory data merely weakened rather than broke epidemiological hypotheses. Thus the overwhelming number of epidemiological studies often led to never-ending controversies about the causes and treatments of diseases, and changing interpretations and explanations were a characteristic result. This is not in any way to deny that all such studies lacked validity. Austin Bradford Hill (an eminent British biostatistician) listed nine criteria that had to be met before a relationship between an agent and a specific disease could be established. His and Richard Doll's famous study of the link between lung cancer and smoking was one such example. Most epidemiological studies, however, were far less rigorous in their methodology.

In many ways the WHI study reflected the uncertainties of the methodology employed by contemporary epidemiology. Recent analyses (in 2009 and 2012) of its finding that a combined HRT treatment (estrogen and progestin) could produce a strong risk of breast cancer identified major shortcomings in this WHI conclusion. First, exposure and outcome were weakly related, if at all. Second, breast-cancer risks increased with advancing age when a female's endogenous estrogen declined, including among women who never took estrogen. Third, most of the correlations published by the WHI were not statistically significant. Other studies could not find a consistently increased risk of breast cancer among those using HRT. Fourth, there was no evidence of a dose-and-duration response; postmenopausal women who had taken conjugated estrogen (a mixture of estrogens) for many years

did not have an increased risk of breast cancer. Fifth, the overwhelming majority of breast-cancer patients never used HRT. Finally, a decline in the incidence of breast cancer began two years
before the WHI report, at a time when HRT was in common use,
thus raising important issues about cause and effect.[17]

The methodological weaknesses of the WHI study were by
no means unique; they were similar to many other modern epidemiological studies that gave rise to murky statistical associations.[18] Nevertheless, the WHI study was quickly followed by a
series of others that seemed to confirm, if not extend, its findings.
In spring 2003, both *JAMA* and the *New England Journal of Medicine* published a series of articles that added to the fears about
estrogen-progestin therapy. These articles raised further questions
about the use of HRT, a therapy that was not directly related to
osteoporosis yet had been a mainstay in its treatment.

In one study published in 2003, examining the relationship
between combined HRT and mental functioning, the investigators found that this therapy increased the risk of probable dementia in postmenopausal women age 65 and over, and it did not prevent mild cognitive impairment in these women. The increased
risk would result in an additional twenty-three cases of dementia
per 10,000 women per year.[19] A second study also found a small
increased risk of clinically meaningful cognitive decline in the estrogen plus progestin group.[20]

The same issue of *JAMA* included articles on the effect of estrogen plus progestin on strokes and breast cancer in healthy postmenopausal women. One article indicated that an excess risk of
ischemic stroke (from an interruption in blood flow to the brain)
was apparent in all age groups; in all categories of baseline stroke
risk; and in women with and without hypertension, a prior history of cardiovascular disease, or prior use of hormones, statins,
or aspirin.[21] A second article found that even short-term estrogen
plus progestin use increased incident (localized) breast cancers—
which were diagnosed at a more advanced stage compared with
those discovered in women taking placebos—and substantially
increased the number of women with abnormal mammograms.[22]

In an accompanying editorial, two physicians at the Feinberg School of Medicine at Northwestern University expressed admiration and approval of the WHI study, which had demonstrated "that alteration of a woman's basic hormonal physiology over decades in the interest of long-term disease prevention is fraught with hazard." The use of combined HRT decreased both the sensitivity and specificity of mammograms because of an increase in radiographic breast density. The therapy thus elevated the risk of developing the breast cancer while simultaneously delaying its detection.[23]

A month later the *New England Journal of Medicine* published a long article detailing the effects of combined HRT on one's health-related quality of life. Randomization for estrogen plus progestin produced no significant effects on a woman's general and mental health, depressive symptoms, or sexual satisfaction. This therapy was associated with a statistically significant but small and not clinically meaningful benefit in terms of improving physical functioning and reducing sleep disturbances and bodily pain after one year of use. Such minor benefits, according to the investigators—limited as they were to one year—did not outweigh the health risks involved in the therapy.[24] Since hormone therapy was linked with decreased risks of colon cancers and hip fracture, were there women who still might benefit from hormonal treatment? In balancing out risks versus benefits, concluded Deborah Grady (from the University of California, San Francisco) in another 2003 article in that journal, the use of HRT for the treatment or prevention of osteoporosis was inappropriate, given the availability of other treatments.[25]

The end of the WHI estrogen-progestin trial in 2002 did not affect the estrogen-only part of the WHI study. In early 2004, however, the NIH also decided to stop the latter trial, which had enrolled 10,739 postmenopausal women, age 50–70, who had had a prior hysterectomy. The data-monitoring board found that estrogen alone indicated no increase in the risk of CHD or colorectal cancer, but there was a 39 percent increase in the risk of strokes and a 34 percent increase in the risk of pulmonary embolisms.

These were counterbalanced by a 23 percent reduction in the risk of breast cancer (which the board deemed statistically not significant) and a 39 percent reduction in the risk of hip fractures. Put in somewhat different terms, there was an absolute excess risk of twelve additional strokes per 10,000 person-years and an absolute risk reduction of six fewer hip fractures. The estimated excess risk for all monitored events in the global risk index was a non-significant two events per 10,000 person-years. Although these data were hardly earth shattering, the board concluded that there was "no overall benefit" in using estrogen for chronic disease prevention. It supported the FDA recommendation that postmenopausal women should use estrogen only for menopausal symptoms "at the smallest effective dose for the shortest possible time."[26]

The findings from the WHI trials had significant implications for osteoporosis therapies. The claims about estrogen-progestin and estrogen-only treatments and their relationships to a variety of chronic diseases raised serious questions. Did reduced fracture rates (assuming such rates were accurate) outweigh the risks of these therapies? Neither the data nor the recommendations resulting from these two WHI trials offered clear answers. In late 2002 the U.S. Preventive Services Task Force of the Agency for Healthcare Research and Quality (AHRQ) recommended against the use of combined estrogen-progestin therapy for preventing cardiovascular disease and other chronic conditions in postmenopausal women. The task force conceded that this therapy could increase BMD and reduce fractures and the risk of colorectal cancer, but concluded that the "use of HRT for primary prevention of chronic conditions requires reevaluation by postmenopausal women and their physicians." Three years later this task force updated its report and recommended against the routine use of HRT for the prevention of any chronic conditions in postmenopausal women, finding that HRT was either ineffective as a preventive treatment or that its harms outweighed its benefits.[27] The unfavorable publicity HRT received had a dramatic impact. In the last quarter of 2003, prescriptions for all forms of hormone therapy fell by 43 percent, and the decline for Prempro alone reached 80 percent. A local San Francisco study of women age 50–74 participating in the Mammography Registry found that between four and six out of ten women stopped taking their hormone pills.[28]

Questioning the use of HRT as a therapy for osteoporosis also began to be more common in the medical literature. The 2004 surgeon general's report on *Bone Health and Osteoporosis* was typical. Postmenopausal hormone therapy, this report noted, had a consistent and favorable effect on BMD at all sites. Research evaluating its impact on fracture rates, however, was more limited. While observational studies suggested that women taking postmenopausal hormones had fewer fractures than those who had never taken hormones, the report stated, such studies were "subject to biases." Women who took hormones, for example, were more likely to take other measures to enhance their health. Up until the mid-1990s, there were few randomized clinical trials of estrogen focusing on fractures as their outcome. But, continued the report, recent meta-analyses of studies that did include fractures as an outcome suggested that estrogen reduced the risk of non-spine fractures by 27 percent and spine fractures by 33 percent. The WHI trials demonstrated the antifracture efficacy of estrogen, but also revealed its deleterious effects.[29]

What, then, was the future for HRT? The 2004 surgeon general's report conceded that estrogens and progestins were approved therapies for relief from hot flashes and symptoms of vulvar and vaginal atrophy. Although effective for the prevention of postmenopausal osteoporosis, these hormone treatments should only be considered for women at significant risk of osteoporosis who could not take non-estrogen medications. The FDA recommendation that estrogens and progestins be used at the lowest possible doses and for the shortest amount of time to achieve treatment goals was inadequate, the surgeon general's report stated, if only because it was unclear whether this advice would lead to long-term benefits in bone health. Many studies indicated that the positive effects of hormone therapy ceased when it was discontinued. Yet, continued the report, despite all of the negatives raised about HRT, it was important to remember that endogenous estrogen was a vital factor in bone health. Hence the surgeon general's report suggested that for women whose bodies, on their own, produced an extremely low amount of estrogen, very low doses of hormone therapy might be a promising treatment.[30]

Thus, in the beginning years of the twenty-first century, HRT for osteoporosis had fallen into disfavor. Its effectiveness and safety had always been a subject of debate, and the publicity that followed the outcome of the WHI study helped to undermine the legitimacy of HRT by raising questions about the safety of this therapy. The statistical analyses purportedly demonstrating the risks of HRT were hardly persuasive, but that fact was all but ignored. The emphasis on relative rather than absolute risk merely stoked public fears. At the same time, the development of other classes of drugs to treat osteoporosis seemed to provide a viable alternative. Taking advantage of the data presented by the WHI trials, pharmaceutical companies were quick to exploit fears about HRT and promote the use of their antiosteoporotic drugs.

The decline in support for estrogen therapy, however, also reflected broader social forces. Older traditions of medical authority had slowly been eroding as younger generations of women began to challenge long-held beliefs about the nature of femininity. The traditional views that women were defined by their sexual organs and that they were inferior to men came under fire by groups seeking equality both in the home and the workplace. Nowhere was this better illustrated than in the changing understanding of menopause. For decades menopause had been equated with physical and mental decline. It was defined as a deficiency disease that required medical treatment. With modifications, such views were expressed in a U.S. House of Representatives hearing in 1991[31] and a Senate hearing the following year. At the latter hearing, Senator Brock Adams (the chairman) expressed a commonly held view. "Menopause and the loss of ovarian hormones plays an important role in the development of diseases and conditions affecting women," he observed in his opening remarks. "Many of the health problems women face, such as heart disease, osteoporosis, urinary incontinence, and cancer, begin in midlife." In subsequent testimony others echoed this theme.[32]

As the role and status of women underwent fundamental changes, however, the very meaning of menopause was also transformed.[33] Nowhere was this transformation better illustrated than in an NIH State-of-the-Science Conference Statement on Management of Menopause-Related Symptoms, held in early 2005. Unlike earlier conferences that were dominated by male physicians, this time women played the leading role. Nine of the twelve members of the panel were women, as were the majority of speakers and those on the planning committee. The opening sentence of the conference statement illustrated the profound change that had occurred. "Menopause is a natural process that occurs in women's lives as part of natural aging." Because women age as they pass from premenopause to postmenopause, the document continued, it was difficult to determine which symptoms happening during this time were due to ovarian aging specifically, and which were due "to general aging and/or life changes commonly experienced in midlife." There was insufficient evidence to conclude that there was a causal relationship between menopause, on the one hand, and difficulty in thinking, forgetfulness, cognitive disturbances, and the prevalence of somatic symptoms, on the other hand. Moreover, existing studies were inadequate for separating the consequences of aging from the effects of menopause. "Menopause," the conference statement concluded, "is 'medicalized' in contemporary U.S. society. There is great need to develop and disseminate information that emphasizes menopause as a normal, healthy phase of women's lives and promotes its demedicalization."[34]

CHANGING THERAPIES

The discord that followed the WHI critique of hormone replacement therapies provided new entrepreneurial opportunities for pharmaceutical firms seeking to expand the market for their anti-osteoporotic drugs. Yet there were those in the research community who expressed concerns about this rising therapeutic enthusiasm.

"We live in an age of large-scale, international, randomized, placebo-controlled trials, mostly financed by the pharmaceutical industry and often designed to introduce a therapy that is novel," wrote B. E. C. Nordin, a pioneer who had been working on the relationship between calcium deficiency and osteoporosis for more than half a century. He observed that the results of these meticulously planned and executed trials financed by pharmaceutical firms were reported in major medical journals and disseminated to the medical profession by persuasive representatives of the companies, thus leading to increased use of the drug under investigation. Nordin, however, had important reservations about them. Treatment with these antiosteoporotic medications was "relatively empirical" rather than "based on subtle scientific indications which could restrict their use." It was also unfortunate, he noted, that few trials compared one treatment with another.

To many family physicians and specialists, Nordin added, bone was "something of a black box," and these clinicians prescribed any therapy that had been shown to increase bone mass, reduce fracture risk, or both. Nordin believed that a word of caution was in order, however, since all cases of osteoporosis did not have the same pathophysiology. When a rise in mean BMD followed a particular therapy, it implicitly concealed the fact that 50 percent of the patients gained bone at a rate above the mean, and 50 percent below, to say nothing about some who lost bone. What were the differences in the metabolic states of those who gained bone and those who lost bone? In 2003, when Nordin's article was published, treatment occurred regardless of a patient's metabolic state. There was also confusion between therapies designed to prevent bone loss and fractures, and those seeking to reverse established osteoporosis. "Low bone density," Nordin noted, "may be the result of rapid bone loss (which requires treatment), but it can equally well be due to low peak bone density, which may not require treatment, or even to small stature, since dual-energy X-ray absorptiometry scanners only give us areal densities which do not fully correct for body size." In his view, what was required was for both clinicians and the pharmaceutical industry to adopt a "more selective, targeted, and sophisticated approach to the treatment of osteoporosis. . . . The present situation is suggestive of an earlier era when all patients with anemia were treated with iron because it had been shown to work in some of them."[35]

Nordin's concerns, however, had virtually no effect. The pace of therapeutic innovation began to accelerate as pharmaceutical firms competed for market shares. The decline in the use of HRT did not in any way shrink the demand for antiosteoporotic drugs; it merely created entrepreneurial opportunities for the introduction of new ones that would presumably prevent osteoporosis and treat individuals with bone diseases.

The publication of the 2004 surgeon general's report on osteoporosis and bone disease was a landmark. Like the surgeon general's report on the dangers of smoking forty years earlier, this later report was designed to galvanize both the medical profession and the public to devote greater attention to osteoporosis. As Tommy G. Thompson (secretary of the Department of Health and Human Services) noted in his introduction, "the prevalence of bone disease and fractures is projected to increase markedly as the population ages. If these predictions come true, bone disease and fractures will have a tremendous negative impact on the future well-being of Americans."[36]

The 2004 report was both a warning about the dangers of osteoporosis and a presentation of opportunities for its prevention and treatment. The report proposed a pyramid approach. The foundation, or base, involved lifestyle changes: proper nutrition (including a diet adequate in calcium and vitamin D), a weight-bearing exercise program, the avoidance of smoking and excessive alcohol consumption, the use of measures to avoid falls, attention paid to diseases that increased the risk of falling, and the avoidance of drugs that caused bone loss or increased the risk of falls. The second tier was the identification and treatment of diseases that produced secondary osteoporosis or aggravated primary osteoporosis. The third tier was pharmacotherapy for those at a high risk for fracture.[37]

Pharmacotherapy included two general categories of drugs.

The first were antiresorptive agents that reduced bone loss, while the second were anabolic agents that built bone. The former included bisphosphonates, estrogen, selective estrogen receptor modulators, and calcitonin. These drugs, the 2004 report noted, should be considered for all patients with osteoporosis. By this date, the FDA had approved two bisphosphonates (alendronate and risedronate) and one selective estrogen receptor modulator (raloxifene). Many estrogen preparations had also been approved, although the surgeon general's report expressed reservations about their use. Anabolic therapy was available for those who still continued to lose bone while on an adequate program of primary prevention and antiresorptive therapy. Only one FDA-approved anabolic agent was available (a synthetic form of parathyroid hormone known as teriparatide). The latter required a daily injection and was far more costly than antiresorptive drugs. The efficacy of bisphosphonates was clear, the report stated; they were also well tolerated (except among those with a history of narrowing or ulcers of the esophagus, or long-standing problems with stomach ulcers and heartburn that required medication). In addition, a variety of antiresorptive and anabolic drugs were in various stages of development, including bisphosphonates that needed to be administered intravenously once a year, thus avoiding the problem of non-compliance.[38]

Three years later the AHRQ issued a report comparing preventive treatments for fractures in men and women with low BMD. Like the surgeon general's report, it maintained that there was good evidence that the bisphosphonates, calcitonin, parathyroid hormone, raloxifene, zoledronic acid, and teriparatide were effective; the evidence on calcium and vitamin D supplements was unclear. The AHRQ analysis did not indicate sharp variations in the efficacy of most of the medications. Instead, a major problem was the low rate of adherence to therapy, which was due in part to the side effects of the medications, an absence of symptoms for osteopenia and osteoporosis, comorbid conditions, age, and socioeconomic status. Few studies evaluated the effect of therapies on men. In 2011 the AHRQ updated its report and paid somewhat greater attention to the side effects of many of the drugs. The report noted that RCTs did not provide data on the frequency with which BMD should be monitored. Moreover, increases in BMD did not predict greater reductions in the risk of fractures. The AHRQ conceded as well that data comparing the effectiveness of short-term versus long-term therapy were not available. Nevertheless, their report found that most of the drugs were beneficial, even though there were variations in their efficacy for different conditions. Overall, the report provided a positive evaluation.[39]

In the opening decade of the twenty-first century, publications dealing with the efficacy, cost-effectiveness, and risks of pharmacotherapy proliferated. Many of the clinical-outcome studies emphasized the benefits of the bisphosphonates, which were considered a first-line therapy for osteoporosis and had the largest base of clinical-trial data. There was general agreement that these drugs increased BMD and reduced the risk of fractures. Yet data on their outcomes were not overwhelmingly favorable. Drugs for the treatment of osteoporosis were generally assessed in three-year trials, with avoidance of vertebral fracture as the primary endpoint (when a trial would be considered successful) and an increase in BMD as a secondary endpoint (an intermediate outcome expected to be achieved during a trial). It was commonly believed that reduced BMD increased the risk of fractures, and that antiresorptive agents increased BMD and would thus led to fewer fractures. Yet an analysis of the data revealed that only a small proportion of the risk reduction for vertebral and nonvertebral fractures observed in the group receiving antiresorptive drug therapy was explained by an increase in BMD.[40] Equally significant, data from the NORA study showed that only 6.4 percent of postmenopausal women with an osteoporotic fracture had T-scores below −2.5. Although fracture rates were highest among this group of women, they experienced only 18 percent of the overall number of osteoporotic fractures; 52 percent of the fractures were among women with a diagnosis of osteopenia (T-scores between −1 and −2.5). Most of the reduction in fracture risks for participants in the trial, therefore, was due to non-BMD determinants of bone strength.[41]

While most studies focusing on the two leading antiresorptive drugs—alendronate and risedronate—reported significant reductions in fracture risks, their data, too, were often less impressive than they appeared at first sight. In one study 5,445 women age 70–79 who had osteoporosis plus a non-skeletal risk factor for hip fractures, and 3,886 women at least 80 years old who had one non-skeletal risk factor for hip fractures or low BMD at the femoral neck, were randomly assigned to two groups. One group was given risedronate, and the other a placebo, for a period of three years. The incidence of hip fractures among those receiving the drug was 2.8 percent, compared with 3.9 percent in the placebo group. The drug treatment was even less effective among the older women selected on the basis of non-skeletal risk factors: the incidence of hip fractures among the treated group was 4.2 percent, compared with 5.1 percent among those assigned to a placebo.[42] According to Nguyen D. Nguyen, John A. Eisman, and Tuan V. Nguyen, who undertook a Bayesian analysis of twelve RCTs between 1990 and 2004, "the efficacy of antiresorptive bisphosphonate therapy on reducing hip fracture is not clear, because evidence from randomized clinical trials is inconclusive." These investigators found that for 18,667 women with low BMD who were either treated with bisphosphonates for one to four years or simply followed during that period, treatment reduced the relative risk of hip fracture by 42 percent. The absolute rate of reduction was fifty-two hip fractures per 10,000 women who treated for three years, a somewhat less impressive statistic.[43]

In general, virtually all of the clinical trials (which were funded by drug companies) reported favorable results. When the American College of Physicians was preparing guidelines on pharmacological treatments for osteoporosis, it analyzed hundreds of randomized controlled trials. This organization found good evidence that alendronate, eridronate, ibandronate, risedronate, calcitonin, parathyroid hormone, and raloxifene, when compared with a placebo, seemed to have prevented vertebral fractures. Evidence from multiple trials and meta-analyses indicated that alendronate and risedronate, compared with a placebo, also seemed to have prevented vertebral and non-vertebral fractures in high-risk populations. In evaluating twelve different drugs, the American College of Physicians found strong evidence that nine of the drugs reduced vertebral fractures, seven reduced non-vertebral fractures, and seven others reduced hip fractures. It conceded that while many of these agents were effective in apparently preventing osteoporotic fractures, the data were insufficient to determine their relative efficacy or safety.[44] The article did not comment, however, on one obvious shortcoming: comparative clinical trials were unlikely to be undertaken unless required as part of the FDA approval process. The continued introduction of new drugs had no obvious advantages over those already on the market, but the former resulted in elevated costs to the consumer. Many clinicians, however, had a preference for new drugs, which they believed were superior to older ones. Moreover, newer drugs, with their relatively long patent periods, were highly profitable to their original manufacturers; older drugs had a limited patent life or, once their patents expired, could be manufactured as generic drugs by firms other than the one that first introduced that drug.

It was clear that every bisphosphonate approved by the FDA increased BMD. It was equally apparent that bisphosphonate therapy reduced vertebral fragility fractures. Yet the meaning of this finding was murky. "There is no consensus on the definition of a vertebral fracture," noted two physicians writing in the *New England Journal of Medicine*. Only one quarter to one third of incidents identified on x-rays as vertebral fractures were found by clinical examinations. As Norton M. Hadler (an authority on musculoskeletal problems) pointed out, a vertebral fragility fracture was generally defined as a 20 percent decrease (2 millimeters) in the height of a vertebra, a finding not related to the general loss in height that is common in the aging process. Nor was there evidence that bisphosphonate therapy decreased the incidence of back pain among those with vertebral fractures.[45]

Most of the results reported by the overwhelming majority of randomized clinical trials of bisphosphonates and other categories of drugs introduced into clinical practice since the mid-1990s— and by 2011 there were more than dozen antiosteoporotic drugs, each having quite different characteristics—were positive.[46] The rule of thumb was to present the findings of these trials in terms of relative risk reduction, a statistic that, on the surface, appeared to suggest that such drugs had powerful positive effects. Yet virtually all of the clinical trials were conducted not by independent investigators, but rather by clinicians subsidized by the pharmaceutical company that manufactured the drug.

The history of zoledronic acid (Zometa), a bisphosphonate manufactured by Novartis, is instructive. Approved by the FDA in early 2002 for the treatment of multiple myeloma and bone metastases from solid tumors, it was subsequently marketed as an antiosteoporotic drug. The HORIZON three-year, double-blind, placebo-controlled trial regarding this drug was conducted at a variety of sites. It enrolled 3,880 patients with a mean age of 73 years, each receiving a fifteen-minute infusion of the drug; an equal number were assigned to receive a placebo. The primary endpoints were the avoidance of new vertebral fractures and hip fractures. Secondary endpoints included improved BMD, bone turnover markers, and safety outcomes. The once-yearly infusion, according to the investigators, "significantly reduced the risk of vertebral, hip, and other fractures," with the risk of vertebral fractures reduced by 70 percent, and hip fractures by 41 percent. The drug, they stated, was also associated with a significant improvement in BMD and bone metabolism markers. Adverse events, including change in renal function, were similar in the two groups, but "serious" atrial fibrillation occurred more frequently in the zoledronic acid group (50 vs. 20 patients).[47]

What, precisely, did the findings of the HORIZON trial signify? The risk of vertebral fractures was reduced from 10.9 percent in the placebo group to 3.3 percent in the zoledronic acid group. The meaning of this statistic, however, is somewhat problematic, given the lack of clarity in the clinical significance of vertebral fractures, as opposed to the more conclusive (and extensive) findings from x-rays. With the clinical data, the 41 percent reduction in the risk of hip fracture (1.4% in the zoledronic acid group vs. 2.5% in the placebo group) meant that 98 persons would require treatment for three years to prevent one hip fracture. But the increase in the risk of serious atrial fibrillation (1.3 in the treated group vs. 0.5 in the placebo group) meant that one person would be harmed among 129 individuals that were treated.[48]

Another randomized, double-blind, placebo-controlled trial of zoledronic acid focused on clinical fractures and mortality after a hip fracture. Two groups of slightly over 1,000 patients each were assigned to receive yearly intravenous doses of the drug or a placebo. The infusions were administered within ninety days after surgical repair of a hip fracture; all enrollees in the trial received supplemental vitamin D and calcium. The median follow-up period was 1.9 years; the endpoint (when treatment stops for an individual enrolled in a trial) was a new clinical fracture. The rates of incurring a new clinical vertebral fracture were 1.7 percent in the treated group and 3.8 percent in the placebo group, a 35 percent risk reduction. In the safety analysis, 101 out of 1,054 patients in the treated group died (9.6%), compared with 141 out of 1,057 (13.3%) in the placebo group, a reduction of 28 percent in deaths from any cause, not necessarily just from a vertebral fracture. Given the mean age of both groups (74.5 years), the researchers conceded that further investigation was needed to more fully understand the reasons why treatment reduced the risk of death, even though they argued that treatment with zoledronic acid after a hip fracture was beneficial.[49] In the same issue of the *New England Journal of Medicine*, Karim A. Calis and Frank Pucino commented that the results of this study appeared "both powerful and compelling." Nevertheless, these two authors stressed that it was essential to obtain additional long-term safety data. Similarly, they noted that other data on the study population—smoking history, degree of mobility, other medications being taken, and the identification of subgroups—were needed to assess how generalizable the outcome was.[50]

The positive findings regarding the efficacy of antiresorptive drugs were generally accompanied by claims that the burdens posed by osteoporotic fractures would continue to rise as
the American population grew older. According to one study, the
medical costs of osteoporosis and fractures in the Medicare population alone amounted to $22 billion in 2008.[51] Given the pro-
jected increase in the proportion of older individuals in the overall population, costs would rise commensurately, thus reinforcing
the belief that only therapeutic interventions could alleviate this
financial burden. Yet there were relatively few studies of the cost-
effectiveness of screening for osteoporosis, or quality-adjusted
life-year analyses of specific therapies. A long 1998 NOF review of
the evidence for the prevention, diagnosis, and treatment of osteoporosis, and a cost-effectiveness analysis of these factors, noted
that it seemed cost-effective to treat one identified category of patients, but not others. Because of the limitations of the data and
the variability of patient preferences, "the lines [between treating
or not treating] must be fuzzy." In general, it was worthwhile to
screen and treat women who had "relatively high near-term risks
of fractures." Nevertheless, the NOF added, a fuzzy line should
not serve as a justification for withholding insurance coverage for
those who seemed, superficially, to be on the wrong side of the
line. The imprecision of the NOF document, however, provided
little that would answer questions about the cost-effectiveness of
interventions.[52]

One study of alendronate therapy for white, early osteopenic
(with T-scores above −2.5), postmenopausal women who did not
have a fracture and did not have additional risks strongly predictive of fractures independent of their BMD, found that the drug
was not cost-effective, even assuming a societal willingness to pay
$50,000 per quality-adjusted life-year gained. The direct medi-
cal costs of pharmacological therapy for such osteopenic women
would be "enormous," with actual costs ranging from $70,000
to $332,000, depending on a person's age and femoral-neck bone
density.[53] In the United Kingdom, the National Institute for
Health and Clinical Excellence (NICE)'s economic appraisals of

the health costs for primary and secondary preventions of osteo-porotic fractures were more restrictive than the guidelines of the Royal College of Physicians. The NICE recommendations set off a sharp debate. Some critics argued that NICE's statements on cost-effectiveness were obsolete, because the cost of alendronate had fallen by about 300 percent when its patent expired. Others claimed that the recommendations ignored the advice of a Guide-line Development Group and the National Osteoporosis Society, which proposed that treatment recommendations be based on a ten-year fracture probability, as they were in the WHO's FRAX algorithm.[54]

Equivocal results of the randomized controlled trials of os-teoporotic drugs notwithstanding, the public and clinicians were besieged by advice proffered by drug company representatives, advertisements in medical journals and on television, and the osteoporotic research community, all of whom emphasized the importance of pharmacotherapy. The NOF was especially active in insisting on the need to both prevent and treat a disease that posed a dire threat to the health and well-being of a population that was growing older.

GUIDELINES FOR SCREENING AND TREATMENT

Given the proliferation of antiosteoporotic medications that claimed to be highly efficacious, questions about screening, the identification of risk factors, and treatment thresholds assumed ever more important roles. A source of concern to clinicians, and especially to a pharmaceutical industry bent on expanding the market for its medications, was that large numbers of women were unaware of or disinterested in their bone status. In the eyes of many clinicians, medical organizations, and the pharmaceuti-cal industry, screening was the most effective tool to demonstrate to women that they needed to undertake measures that would either treat or prevent osteoporosis.

During the 1980s and 1990s, there had been many debates about identifying the most vulnerable groups who could best benefit from periodic screenings and treatment. For a variety of reasons a general consensus did not emerge. The proliferation of drug therapies, however, proved to be crucial in altering the situation. What was the use, after all, of developing effective medications if women remained oblivious to their advantages? During the first decade of the twenty-first century, therefore, pressure for osteoporosis screening mounted. The result was the proliferation of clinical-practice guidelines and recommendations designed to identify those groups in need of screening. Screening would involve measuring BMD and, at the same time, identifying risk factors that increased vulnerability. Both would contribute to the goal of maintaining bone health and preventing vertebral and hip fractures.

Numerous public and private agencies had issued guidelines for the management of osteoporosis and screenings for this disease before 2000. Yet profound differences between them were common. One study that evaluated twenty-one guidelines issued between 1998 and 2001 (eight by medical societies, six by national groups, six by government agencies, and one by an international group) found that half were not suitable for use in practice. The guidelines' methodological quality was low, and virtually none covered dissemination issues, although a few were judged to be acceptable. Nearly a third of the guidelines acknowledged the involvement of or support from the pharmaceutical industry: 87 percent of their authors had some form of interaction with the industry, and 58 percent received financial support for their research. "Industry has a vested interest in supporting the recommendations of the osteoporosis guidelines," the authors of the evaluative study warned, "which raises concerns about industry's potential to influence the guideline development process."[55]

After 2000, screening recommendations issued by various government agencies and professional organizations proliferated. In 2002 the U.S. Preventive Services Task Force (USPSTF) issued guidelines on screening for postmenopausal osteoporosis. Its recommendations were based on a variety of literature searches covering more than three decades, as well as on the views of experts. The USPSTF could identify no studies about the effectiveness of screening in reducing osteoporotic fractures. Hence recommendations had to rely on information from risk-factor assessments or BMD testing as a means of adequately identifying women who could benefit from treatment. After an extensive review of the existing evidence, the agency recommended that routine screening begin at age 60 for women at increased risk for osteoporotic fracture (based on a list of about twenty risk factors). It made no recommendation for or against routine screening for postmenopausal women younger than 60, or women age 60–64 who were not at risk for osteoporotic fractures. The USPSTF estimates were based on statements about the benefits of detecting and treating osteoporosis that were found in studies of bisphosphonates (although the agency noted that other treatment options were available).[56]

Nine years later, the USPSTF went much further. It recommended screening for osteoporosis in all women age 65 years or older, and in younger women whose fracture risk was equal to or greater than that of a 65-year-old white woman who had no additional risk factors. The USPSTF conceded that evidence was lacking about optimal intervals for repeated screening, and it could offer no recommendation for screening men who had no previously known fractures or secondary causes of osteoporosis. The USPSTF found no evidence that screening posed potential harms.[57]

The International Society for Clinical Densitometry (publisher of the *Journal of Clinical Densitometry*) was among the most active organizations in urging expanded BMD testing. Founded in 1993 by a group of specialists, the organization was dedicated to bringing bone densitometry into wider clinical practice, increasing public awareness of it, and encouraging its use. Between 2001 and 2003, the society's official position was that BMD testing should be used for all women age 65 and older, all men age 70 and older, and adults under age 65 with a variety of risk factors. During the remainder of the decade, this organization played an important role in making its standards the prevailing norm.[58]

Nowhere was the rising preoccupation with screening for osteoporosis better illustrated than in the WHO's creation of Fracture

Risk Assessment Tool (FRAX) in 2008. FRAX was developed under the direction of John Kanis, with the support of many individuals and organizations, such as the American Society for Bone and Mineral Research, the NOF, the International Society for Clinical Densitometry, and the International Osteoporosis Foundation. FRAX was a web-based algorithm designed to calculate the ten-year probability of major osteoporosis-related fractures in men and women, based on easily obtained clinical risk factors and optional use of BMD screenings of the femoral neck. Its design was intended to avoid the shortcomings of the BMD screening guidelines, which often identified not only those at high risk for fractures as candidates for therapy, but also included those with low BMD who were not at risk.

The FRAX model used data from nine population-based cohorts from around the world, and that information was confirmed by data from twelve prospective studies. Fracture probability was calculated by using age, body mass index, sex, ethnicity (U.S. only), prior fragility factors, a family history of osteoporosis, smoking, glucocorticoid use, rheumatoid arthritis, alcohol use, and secondary osteoporosis (e.g., from type 1 diabetes, chronic liver disease, etc.). FRAX was designed to be used when there was uncertainty about when to begin treatment, for patient selection in clinical trials, and for periodic measurements of fracture risks over time. It was not intended for patients undergoing treatment, for individuals for whom treatment was clearly indicated, or for patients with T-scores above –1.0. FRAX had its limitations, however. For example, dose response was not taken into account for several risk factors (smoking, alcohol, and the number and severity of fractures). Nevertheless, according to Nelson B. Watts, in discussing its applications in clinical practice, FRAX was "a major achievement for the medical community and is currently the gold standard among fracture models."[59]

At the same time, the NOF brought out the third (2008) edition of its *Physician's Guide*. The second edition, five years earlier, had focused solely on postmenopausal Caucasian women. Treatment thresholds in the 2003 edition were individuals with a previous hip or vertebral fracture; those with a T-score below −2.0 at the hip; and those with low BMD (a T-score of −1.5, and −2.0 at the hip) who also had such clinical risk factors as a family history of osteoporosis, a history of low-trauma fracture, smoking, and low body weight. The 2003 NOF guidelines went beyond the WHO threshold, as the latter defined osteoporosis in terms of a T-score of −2.5 or below. NOF's rationale for expanding the definition was based on the fact that the majority of hip fractures were among women with T-scores above −2.5. Nonetheless, the 2003 NOF guidelines were somewhat limited in scope and did not assess the relative strengths of clinical risk factors.[60]

The 2008 edition of NOF's *Physician's Guide* was not simply an updated version of the two previous editions, however. It dramatically expanded osteoporosis screening and treatment thresholds to include virtually the entire population of older females and males. "Osteoporosis," according to the 2008 *Guide*, "affects an enormous number of people of both sexes and all races, and its prevalence will increase as the population ages." Using data from the National Health and Nutrition Examination Survey III (NHANES III), the 2008 NOF *Guide* estimated that more than 10 million Americans had osteoporosis, and an additional 33.6 million had low bone density in their hips. According to NOF's figures, one in two Caucasian women, and one in five men, would experience an osteoporosis-related fracture in their lifetimes, with most of these fractures due to low bone mass. Moreover, osteoporosis-related fractures created a heavy economic burden, since they resulted in 432,000 hospital admissions, 2.5 million medical office visits, and about 180,000 nursing home admissions annually. The costs of osteoporosis-related fractures had been estimated at $17 billion for 2005, and NOF warned that this figure that could double or triple by 2040.[61]

In NOF's 2008 *Guide*, many factors were associated with an increased risk of osteoporosis-related fractures. Multiple lifestyle elements played a role: inadequate calcium and vitamin D intake; high intakes of caffeine, salt, and aluminum; excessive alcohol consumption; smoking; falling; bodily thinness; inadequate physical activity or immobilization; genetic factors; hypogonadal states; endocrine, gastrointestinal, and blood disorders; rheumatic, autoimmune, and similar diseases; and a variety of miscellaneous physical conditions. Finally, some medications (e.g., glucocorticoids, anticoagulants, etc.) contributed to osteoporosis.[62]

Since the majority of osteoporotic fractures resulted from falls, NOF signaled the importance of evaluating risk factors for falling. Environmental risk factors included household obstacles, poor lighting, loose throw rugs, a lack of assistive devices in the bathroom, and slippery outdoor conditions. Many medical risk factors were also consequential: age; anxiety, depression and agitation; arrhythmias; dehydration; female gender; malnutrition; medications causing oversedation; poor vision, along with the use of bifocals (or variants); and impaired mobility. Finally, neurological and musculoskeletal risk factors played a role, such as kyphosis (outward curvature of the spine), poor balance, reduced proprioception (response of nerve endings to stimuli related to position, posture, equilibrium, or an internal condition in the body), and weak muscles. Many these risk factors were included in the WHO FRAX model.[63]

In addition to a physical examination, the NOF advised clinicians to gather a patient's complete medical history, since metabolic bone diseases such as hyperparathyroidism or osteomalacia might be associated with low BMD and thus could be treated with specific therapies. A diagnosis of osteoporosis, however, was to be established by measuring BMD, although a clinical diagnosis could be made for at-risk individuals who sustained a low-trauma fracture. Who, then, should be tested for BMD? In dealing with this question, the NOF, echoing the position of the International Society for Clinical Densitometry, offered one of the most expansive recommendations to date. The 2008 NOF *Guide* recommended BMD testing for several categories: all women age 65 and older and all men age 70 and older, regardless of clinical risk factors; postmenopausal women and men age 50–69, based on their clinical risk-factor profile; women transitioning into postmenopause, if they had a specific risk factor associated with increased fracture risks; adults who experienced a fracture after age 50; adults either with certain conditions (e.g., rheumatoid arthritis) or taking a medication causing bone loss; anyone being considered for pharmacological therapy for osteoporosis; anyone being treated for osteoporosis, so as to monitor the treatment's effect; and anyone not receiving therapy in whom evidence of bone loss would eventually lead to treatment.[64] A year later, the American College of Preventive Medicine issued essentially the same guidelines for BMD screening. It recommended that all adults age 50 and over be evaluated for risk factors related to osteoporosis, and that women and men age 50–69 should undergo screening if they had one major or two minor risk factors. In addition, women over the age of 65 and men over the age of 70 should universally be screened.[65]

The 2008 NOF *Guide*'s recommendations for pharmacological treatments for osteoporosis were equally as expansive as its screening recommendations. Postmenopausal women and men age 50 and older should be considered for treatment if they fell into one or more of three categories: a hip fracture or a vertebral fracture; a T-score below −2.5 at the femoral neck or spine; and low bone mass (T-scores between −1.0 and −2.5 at the femoral neck or spine) accompanied by a ten-year probability of a hip fracture greater than 3 percent or a ten-year probability of a major osteoporosis-related fracture greater than 20 percent, based on the United States' adoption of the WHO algorithm.[66]

Aside from its extensive list of individuals who would benefit from BMD screening, the NOF's 2008 *Physician's Guide*—following a developing trend—contributed to the growing consensus that osteoporosis was a disease that cut across gender lines. In a certain sense the United States lagged behind other European nations, which had begun somewhat earlier to identify osteoporosis as a problem affecting both males and females. For decades in the United States, osteoporosis and menopause had (for all intents) been synonymous and reflected the belief that females were weaker, more vulnerable, and inferior to their male counterparts. The feminist attack on the traditional interpretation of menopause, the demands for full equality between the sexes, and the growing number of women in medicine and other scientific professions slowly began to create pressure for change. If menopause was simply a stage in a woman's life, it became increasingly difficult to identify osteoporosis solely as a female malady. Did not males experience fractures as well, even if their fracture rate was less than that of females? Under these circumstances, the stereotype of osteoporosis as simply a disease of postmenopausal women began to weaken. The 2008 NOF *Guide*'s recommendations on BMD screening reflected this changing view of osteoporosis.

Nowhere was the desexualization of osteoporosis better illustrated than in the medical literature of the early twenty-first century. "Osteoporosis in men," noted three members of the Mount Sinai Bone Program at the Mount Sinai School of Medicine in New York City, "is an overlooked yet increasingly important clinical problem that, historically, has not received the same degree of awareness as with women." The authors noted that males had a lower risk of fractures, since aging caused less microstructural damage in men, and they had beneficial geometric adaptations that led to stronger bones, compared with women. Nevertheless, the increased morbidity and mortality associated with fractures in men made osteoporosis a significant public health burden.[67]

Feminist critiques of the medicalization of older women's bodies led to attacks on the concept of the "weak woman," even though older women were at greater risk for fractures than men, Samantha L. Solimeo observed. But it was precisely the sexing of osteoporosis that gave men a false sense of security, decreased their access to treatment, and led to poorer outcomes. Solimeo believed that research and interventions had to take into account feelings of masculinity that led to feelings of emasculation when men were diagnosed with a "woman's disease."[68] Slowly but surely, the number of epidemiological studies of BMD among men and the need for screening and treatment for men and women began to proliferate.[69]

The 2003 and especially the 2008 editions of the NOF's *Physician's Guide* had major implications. The former went beyond the WHO definition of osteoporosis as a T-score below −2.5 and recommended treating all women with T-scores between −1.5 and −2.0 if they had one or more risk factors for osteoporosis. The 2008 edition expanded NOF's screening guidelines to include virtually all older persons.

What were the potential impacts of these changing treatment and screening thresholds? In an analysis of the 2003 *Physician's Guide* recommendations, four Dartmouth Medical School faculty members found that the new threshold increased the number of individuals requiring treatment from 6.4 to 10.8 million among women 65 and older (at a net cost of at least $28 billion), and from 1.6 million to 4.0 million among women age 50–64 (at a net cost of $18 billion). Focusing on hip fractures, the investigators noted that there was little evidence that treating an additional number of women in these age categories would reduce the incidence of such fractures. Although BMD was easy to quantify, it was a relatively weak risk factor, and most hip fractures occurred in women who were not diagnosed as having osteoporosis. "Our findings," the authors concluded, "also highlight the general problems raised by expanding disease definitions, which always mean that the number of affected people rises (and the market for treatment is expanded)." The Dartmouth investigators expressed concern that the WHO study group responsible for the current definition of osteoporosis had been funded by three drug companies (Rorer, Sandoz, and SmithKline Beecham), and that the NOF and the American College of Obstetricians and Gynecologists (whose guidelines were substantially the same as those of the NOF) received support from the pharmaceutical industry. "We do not know that this funding influences their decision making about treatment threshold for osteoporosis; however, their relationship with industry is a financial conflict of interest. In our opinion, organizations with such conflicts should not be involved in creating or expanding disease definitions."[70]

The potential impacts of the NOF's 2008 recommendations were equally substantial. One investigation applied data from the Study of Osteoporotic Fractures (SOF) to the 2008 NOF treatment guidelines, in order to estimate the proportion of older white women who would be recommended for pharmacological treatment. SOF was a community-based sample of women from four urban communities, a group that was comparable with the cohort of white females more than 65 years old who participated in the NHANES III study. The results were astonishing. Applying the NOF guidelines to the SOF data, an estimated 72 percent (at a minimum) of white women over 65, and 93 percent of those over 75, would be recommended for treatment. When drug treatments for osteoporosis were advised for this great a proportion of older women, the authors noted, it was important that the assumptions that underlay the analysis be based on robust evidence. A trial of bisphosphonate or similar drug treatments in women and men with just "low bone mass" would be informative.[71]

Another analysis of the NOF guidelines raised other compelling issues. Investigators applied these guidelines to the Framingham Osteoporosis Study, composed of participants from two cohorts of the famous Framingham Heart Study. The Framingham Osteoporosis Study, conducted from 1987 to 2001, included 1,946 postmenopausal women and 1,681 men over the age of 50 who had information available on their BMD. The osteoporosis-study researchers found that nearly half of the Caucasian women and one-sixth of the men age 50 and older would be recommended for treatment if the NOF guidelines were followed. Extrapolated to the national level, 19.5 million Caucasians would be receiving drug treatment. Since the population of men and women greater than 65 years old was expected to double by 2030, the number of persons recommended for drug treatment would exceed 40 million within two decades. The economic consequences were substantial, and Framingham authors suggested that the 2008 NOF guidelines "may need to be re-evaluated with respect to budget impact."[72]

Other studies came to somewhat more favorable conclusions. An analysis of the U.S. population in NHANES III described the prevalence of risk factors used in the FRAX-based NOF *Physician's Guide*. This analysis found that 20 percent of the men and

37 percent of the non-Hispanic white women age 50 and older were at sufficient risk for fractures to warrant pharmacotherapy to lower that fracture risk.[73] A similar study using NHANES 2005–2008 data and applying the same NOF guidelines concluded that 19 percent of the men and 30 percent of the women age 50 years and over warranted consideration for pharmacotherapy. Of special concern was the high prevalence of low BMD and osteoporosis in the Mexican American population, which exceeded the levels among non-Hispanic whites.[74] Another study using NHANES III data found a significant risk of hip fracture in adults 65 or more years old who, according to NOF guidelines, were candidates for treatment to lower their fracture risk.[75] These analyses indicated that the ability to predict fracture risk in those who were 50–65 years old and in non-whites was another aspect that needed to be addressed.

The NOF was not alone in developing guidelines for osteoporosis screening and treatment. In 2008 the American College of Physicians gave its members two strong recommendations. First, clinicians should offer pharmacological treatment to women and men known to have osteoporosis, and to those who experienced fragility fractures. Second, clinicians should choose among the various pharmacological treatment options on the basis of the risks and benefits a particular treatment would offer to a specific patient. Pharmacological treatment for those categorized simply as being at risk for developing osteoporosis, however, received only a weak endorsement.[76]

In a similar vein, the North American Menopause Society provided guidelines for its members in 2006 and 2010. The former version supported BMD testing for all women age 65 and over, plus those under 65 who had certain risk factors. It also suggested pharmacological treatments for women with T-scores between –2 and –2.5 who had concurrent risk factors. The 2010 guidelines incorporated the FRAX model, which relied on multiple risk factors to identify those women with low bone mass who needed treatment. What was especially noteworthy in 2010 was the friendlier attitude manifested by the organization toward HRT.

At the Thirteenth World Congress on Menopause, held in Rome in 2011, the International Menopause Society issued a statement on HRT, noting that healthy women under 60 years of age ought not be concerned about its safety profile. Although HRT should not be prescribed according to an all-embracing formula, it benefits—including fracture prevention—could not be ignored.[77]

RISK FACTORS AND HIP FRACTURES

By the beginning of the twenty-first century, the focus had clearly shifted from the prevention of vertebral fractures to the prevention of hip fractures. The prevailing beliefs were that risk factors were well known and that screening could identify those at risk for hip fractures. Treatment—largely pharmacological—would follow, thus reducing the burden on both individuals and society.

Yet the concept of risk, the prevalence of hip fractures, and their relationship to the diagnostic category of osteoporosis were at best obscure. If the overwhelming majority of hip fractures were found among individuals with low BMD (a T-score of −2.5), it was then plausible to argue that such fractures were a result of the disease. Such, however, was not the case. L. Joseph Melton III Melton and his colleagues—using the Delphi method, which relied on expert group judgment in the absence of adequate data—concluded that osteoporosis was responsible for 90 percent of all hip fractures.[78] The SOF study, which followed 9,704 Caucasian women age 65 and older for nearly a decade, had quite different findings. While the results showed that almost all types of fractures had an increased incidence in women with low BMD, the proportion of fractures attributable to osteoporosis (based on its standard definition, a T-score below −2.5) was at best modest, ranging from less than 10 percent to 44 percent for specific types of fractures, to around 15 percent for all types of fractures combined. About 28 percent of the hip fracture cases were attributable to osteoporosis, defined using total hip BMD. Although conceding that the importance of low BMD on the risk of fractures should not be underestimated, the investigators concluded that there must be other, equally important factors contributing to those risks, including falls.[79]

Most hip fractures were due to falls, and they occurred among individuals not meeting the standard definition of osteoporosis.[80] Under these circumstances, pharmacological treatment would not prevent more fractures, if only because such therapy would not reduce the risk of falling. How common were falls among aged persons and what factors were responsible for falling? A 2012 statement form the U.S. Preventive Services Task Force noted that falls were the leading cause of injury among adults 65 and older and accounted for substantial medical costs. Perhaps 30–40 percent of community-dwelling adults age 65 or more fell at least once per year.[81] An evaluation of thirty-two studies found that mean costs ranged from $2,044 to $25,955 per fall victim, $1,059 to $10,913 per fall, and $5,654 to $42,840 per fall-related hospitalization, depending on the severity of the fall.[82] While falls had multiple causes, several studies found a consistent association between falls in older people and the use of most classes of psychotropic drugs, including tricyclic antidepressants and selective serotonin-reuptake inhibitors. Similarly, those taking more than three or four medications were also at risk for recurrent falls. Inappropriate medication use among elderly persons residing in the community was common.[83] Eyesight provides a reference frame for postural balance and stability, so visual impairment was also strongly linked to an increased risk of fractures. Cataracts were the most common cause of vision-related fractures, and those individuals who had undergone cataract surgery had lower odds of hip fractures.[84] Other health problems that were common to the elderly also enhanced the odds of falling, such as impaired gait and balance, cardiovascular problems, and cognitive deficits.[85]

Organizational clinical guidelines emphasized the importance of fall prevention among older persons, such as the 2001 statement issued by the American Geriatrics Society.[86] The society's revised 2010 guidelines included a variety of interventions, including a customized exercise program, mitigation of risk factors for falls in the home, cataract surgery for impaired vision, a reduction in or withdrawal of medications, treatment for postural hypotension, management of heart rate and heart rhythm abnormalities, and vitamin D supplementation. The society recommended that screening for falls (or the risk of falling) be performed by qualified health care providers.[87]

Several federal agencies issued their own guidelines. The CDC's 2008 guidelines and the 2010 revision recommended three categories of intervention: exercise, home modifications to reduce hazards, and multifaceted screening that included a vision exam and medication review. The National Institute on Aging offered similar recommendations. The USPSTF updated its 1996 recommendations in 2012 and supported physical exercise or physical therapy plus vitamin D supplementation for community-dwelling adults age 65 or older who were at increased risk for falls. The USPSTF did not endorse multifactorial risk assessment, which was judged to have only a small net benefit.[88]

Compared with pharmacotherapy, fall reduction programs did not receive much attention, although it was clear that fall risks were a major contributor to the risk of fractures. "We conclude," noted a group of investigators studying the pattern of falls, "that there may be a substantial 'gap in care' for community-dwelling older fallers who are discharged home" after an emergency-room admittance.[89] Few general practitioners, or even specialists, assessed the risk of falling among their older patients, and the subject was also overlooked in many of the professional publications dealing with fractures.[90]

CONTRAINDICATIONS

Despite the heightened importance accorded to the prevention and treatment of osteoporosis by the medical profession and the pharmaceutical industry, to say nothing about predictions of its increasing prevalence as the American population aged, a series of disquieting developments began to challenge long-held beliefs. The claim that pharmacotherapy was safe, the belief that the prevalence of osteoporosis and the number of hip fractures would rise as the number of elderly persons increased, and the view that as women became educated about the dangers of osteoporosis they would follow medical advice, all began to come under closer scrutiny as new data arose and critics challenged these beliefs. Although its medical legitimacy was not threatened, the diagnosis of osteoporosis, and its preventive and treatment modalities, became somewhat more problematic.

Unlike HRT, the relative safety of bisphosphonates seemed to be confirmed by numerous clinical trials since their introduction in the 1990s. But clinical trials, however significant, are of relatively short duration, and many adverse effects often did not manifest themselves until the passage of time. It turned out that while bisphosphonates increased bone mass, they also inhibited bone turnover, which led to increased mineralization, resulting in brittle bones. A group of oral surgeons reported that in their practice, between 2001 and 2003, they had identified sixty-three cases of patients with necrotic lesions in the jaw and a history of chronic bisphosphonate therapy.[91] Other investigators found cases of osteonecrosis of the jaw in connection with the use of bisphosphonates for managing metastatic disease to the bone and its resultant osteoporosis. In treatments for myelomas and breast cancers, 10 percent of the patients receiving zoledronic acid and 4 percent receiving pamidronate developed this jaw condition.[92] Another small study of nine patients treated with alendronate—either alone or in combination with estrogen—for three to eight years found a potential risk of oversuppressing bone turnover, which impaired some of the biomechanical properties of bone.[93] In an editorial, Susan M. Ott noted that "the profound suppression of bone formation could have negative effects that occur after long-term accumulation in the skeleton. There is no good surrogate for the passage of time, and it will be at least a decade before we have any data about whether the potential negative effects would ever predominate over the known positive effects."[94]

The increasing recognition that the use of bisphosphonates was linked with osteonecrosis of the jaw led the American Society for Bone and Mineral Research to appoint a task force in 2007 to address the problem and offer recommendations. The group noted that the risk of bisphosphonate-associated osteonecrosis of the jaw was relatively low: 1 in 10,000 individuals, and less than 1 in 100,000 patient-treatment years, although the risk was far higher among cancer patients. The society recommended that an external agency follow up and validate reports of osteonecrosis of the jaw, in order to have more accurate information. The group also identified several clinical and basic research questions that would lead to the formulation of a research agenda to better understand, prevent, and treat this condition.[95]

At about the same time, isolated reports of subtrochanteric femur fractures appeared. The earliest was a retrospective review of the operating records of all orthopedic surgeons at two Singapore hospitals whose patients who had undergone surgery for a femur fracture between May 2005 and February 2006. Nine out of the thirteen women identified by the review had been on alendronate for at least three years. Since it was unusual for patients receiving alendronate to sustain this type of fracture, the group suggested that its findings identified "a potential, originally unrecognized, side effect of prolonged pharmacological suppression of bone turnover."[96] In 2008 and 2009, other articles based on either a single or a relatively small number of cases came to similar conclusions.[97] The paucity of such cases, and the fact that there were no population studies comparing the incidence of subtrochanteric fractures among people treated with alendronate versus a group with comparable bone density who had not been treated with this drug, created uncertainty.[98] Two studies in 2010 (both of which received funding from pharmaceutical firms) that attempted to evaluate the claim of a relationship between femur fractures and alendronate use came to slightly different conclusions. A secondary analysis of three bisphosphonate trials (FIT, FLEX, and HORIZON) revealed that fractures of the subtrochanteric or diaphyseal femur were "very rare, even among women who had been treated with the bisphosphonates for as long as 10 years."[99] A Canadian study of 52,595 women age 68 and older who had been treated with a bisphosphonate for five years or longer had an increased risk of fractures, but the absolute risk was low. A follow-up of the study's participants recorded 71 fractures (0.13%) in the subsequent year and 117 (0.22%) after two years.[100] Danish investigators, using an age- and gender-matched cohort assembled from national healthcare data found a higher risk of hip and femur fractures among alendronate-treated patients, compared with the control group, but large cumulative doses of alendronate were not associated with a greater fracture risk than small cumulative doses, thus "suggesting that these fractures could be due to osteoporosis rather than to alendronate."[101] A cohort study, using a large American database that studied fragility fractures from 2000 to 2009 at all sites on the femur after up to eight years of therapy with alendronate or risedronate, did not find a reversal of fracture protection with long-term use of these drugs.[102]

The controversy over the relationship between oral bisphosphonates and atypical subtrochanteric femur fractures led the FDA to issue a drug-safety communication, announcing an ongoing review of this issue and urging health care professionals to be aware of such fractures in patients taking bisphosphonates. In 2012 the FDA review (which was based on three long-term extension trials in which the duration of treatment ranged from six to ten years)[103] came to somewhat equivocal conclusions. The FDA noted that there was a potential for cumulative risk in the use of bisphosphonates, although the agency's report did not list specific risks. It focused instead on how long patients should be on a drug regimen. Those with a low risk of fracture whose bone density approached normal were candidates for ceasing drug therapy after three to five years. Older patients with a history of fracture and a BMD in the osteoporotic range might benefit from continued bisphosphonate therapy. The FDA review recommended that decisions to continue treatment "must be based on individual assessment of risks and benefits and on patient preference," although this statement offered little specific guidance.[104]

In the same issue of the *New England Journal of Medicine* in which the FDA review appeared, a group of five well-known authorities extrapolated from the same data used by the FDA and recommended that patients with low BMD at the femoral neck (a T-score below –2.5) could benefit from a continuation of bisphosphonate therapy; patients with existing vertebral fractures and a T-score between –2.0 and –2.5 might benefit; and those with a T-score above –2.0 and a low risk of vertebral fractures were unlikely to benefit. The investigators conceded, however, that additional data might alter their recommendations.[105]

The FDA recommendations did not focus on specific adverse reactions to bisphosphonate therapy, but they did suggest that concern over the use of these drugs was mounting, even in the relative absence of data on the incidence and prevalence of adverse events. In addition, the number of legal suits against pharmaceutical companies by individuals who claimed that they were harmed began to increase.

Under these circumstances it was not surprising that osteoporosis researchers and clinicians began to examine the benefits and risks of bisphosphonate therapy. There was little disposition to question the belief that such drug treatments prevented some types of fractures. "We know why to treat—fractures are associated with morbidity, mortality, and cost," Ego Seeman observed. "We know who and when to treat—those at high absolute risk for fracture. We know what drugs to use—those tested in credibly designed and executed trials." The remaining questions were to determine how long treatment should be sustained, and whether prolonged therapy resulted in harm. Were benefits achieved in the first three years of treatment lost on its cessation? Admitting that the evidence to answer such questions was limited, Seeman nevertheless maintained that "stopping therapy is more likely to do harm than continuing therapy."[106]

Bisphosphonates worked by inhibiting bone resorption, which in turn increased bone mass, since a smaller amount of old bone was removed. It was biologically plausible that atypical fragility fractures could result from drug therapies that suppressed bone remodeling. The difficulty, most researchers conceded, was the absence of high-quality research proving causality. There were no studies comparing antifracture benefits for those who discontinued a drug after three to five years with those who continued treatment. Given that adverse events were rare, there was a general consensus that the benefits of therapy far outweighed the risks. As one authority asserted, even if the risks of bisphosphonate therapy were real, "it is important that the major therapeutic benefits that can accrue from bisphosphonates' appropriate targeted use are not lost as a result of the anxiety concerning these extremely rare adverse events."[107] The weakness or absence of longitudinal data, however, led many to advise caution in the use of bisphosphonates and encourage a review of a patient's BMD and risk profile after three to five years of therapy, in order to avoid treatment for those at low risk.[108]

Concern with the negative consequences of antiresorptive therapy led John C. Stevenson (of the National Heart and Lung Institute at the Imperial College in London) to call for a reassessment of HRT therapy. In a lecture at the annual meeting of the British Menopause Society, Stevenson pointed to safety concerns associated with the increasing use of bisphosphonates. Moreover, their skeletal retention times often lasted for many years, with unknown consequences. Challenging the findings of the WHI about HRT, Stevenson insisted that it be rehabilitated as a first-line therapy for the prevention of postmenopausal osteoporosis, since HRT was effective, cheap, and as safe as other medications.[109]

At the same time that pharmacotherapy had become central to the treatment and prevention of osteoporosis, a series of epidemiological studies posed troubling questions about future developments. For decades the osteoporosis community had been warning that the burden of osteoporosis-related fractures, along with their costs, would continue to increase as the number of older persons susceptible to falls and fractures rose. One group of researchers found that in 2005 there were more than 2 million fractures, leading to direct medical costs of $17 billion. "By 2025, annual fractures and costs are projected to grow by 50% and will surpass 3 million and $25 billion, respectively."[110]

Yet demographic data from several nations contradicted this projection. In Finland, for example, there was a rise in the incidence of age-adjusted hip fractures from 1970 to 1997. Between 1997 and 2004, however, the trend was reversed. Among women the decline was from 494 hip fractures per 100,000 persons in 1997 to 412 in 2004, and among men, from 238 to 223. Similar trends were found in Scandinavia. Possible explanations included a healthier older population, a rise in body weight and BMD, and improved functional ability among the aged. Preventive measures—HRT, smoking cessation, calcium and vitamin D supplements, and bone medications—at best played minor roles. The exact reason for the change in hip fracture incidence, the investigators concluded, was unknown.[111]

Trends in Finland and Scandinavia were by no means unique. In Rochester, Minnesota, age-adjusted hip fracture incidence rates for women increased between 1928 and 1950, but then declined. Between 1980 and 2006, the rate for women fell by 1.37 percent per year, and for men, by 0.06 percent. The decline could not be attributed to drug therapies, because a minority of the women and none of the men were treated with bisphosphonates.[112] A Swiss study of hip fractures between 1991 and 2000 also found a decline in these fractures, despite an increase in the population at risk. Again, there was no clear explanation why. Drugs were not a factor, because of the low prevalence of their use and the fact that a general awareness of osteoporosis and its consequences was low.[113] Much the same situation was evident in the Netherlands.[114]

One of the largest analyses of trends in hip fracture rates was conducted in Canada, covering over 570,000 hospitalizations for hip fractures between 1985 and 2005. This study found a decrease in age-adjusted hip fractures of 1.2 percent per year between 1985 and 1996, and a 2.4 percent decline per year from 1996 to 2005. The rate of decrease occurred in both sexes, but was slightly higher among females. Once more, the investigating group conceded that they could not isolate factors to explain this decline. BMD testing and pharmacotherapy might have been a factor, but low treatment rates among females and the absence of treatment among males vitiated such an explanation. There was also little evidence to suggest that declining smoking rates and improvements in physical activity, calcium intake, and bodily vitamin D status could account for the magnitude of the reductions in hip fractures. While decreasing incidence rates were not grounds for complacency toward osteoporosis prevention and treatment, the authors emphasized that overestimations of future fracture rates had implications for health care administrators responsible for coordinating and prioritizing health care delivery.[115]

The unexplained decline in the incidence of hip fractures among both women and men who had never received medical testing or treatment for this condition posed other dilemmas. Under such circumstances, how could recommendations for universal screenings for osteoporosis—and pharmacotherapy for postmenopausal women and elderly men—be justified, particularly in view of rising costs? Although the changing epidemiology of hip fractures remained somewhat of a mystery, as yet there was no widespread disposition to modify either the diagnostic category of osteoporosis or the prevailing standards of screening and treatment. The authors of an Australian study that found a decreasing trend in the number of fall-related hip fractures in Victoria nevertheless insisted that in the future, the burden of fall-related hip fractures would escalate, due to the aging of the population.[116]

Pharmaceutical companies not only continued their marketing activities, but also pursued studies of physicians' prescribing practices, in the hope of modifying these practices and ensuring that doctors were aware of the benefits of pharmacotherapy. Amgen, for example, created the POSSIBLE US Treatment Registry Study, a large prospective listing of postmenopausal women receiving treatment for osteoporosis. A group of investigators, including two employees of Amgen, analyzed data from family-practice physicians, internists, gynecologists, and 4,917 women (postmenopausal for at least one year and initiating, switching, or augmenting therapy) who were enrolled in the registry for more than two years. The study found that variations in care existed and that patients at high risk for fractures were, in the authors' opinion, not always appropriately managed. Younger physicians were more likely than their older counterparts to utilize DXA to monitor patients and prescribe bisphosphonates. Thus the study noted that there was an opportunity for quality-improvement interventions by educating physicians who had been in practice for long periods of time, so they became aware of advancements in medical knowledge and changes in treatment options. It was clear that the study (which was supported by Amgen) was intended to enhance physicians' awareness of osteoporosis and thus make them receptive to prescribing active therapeutic regimens (drugs).[117]

The effort to educate physicians about the dangers of osteoporosis was only one of the issues. Equally significant was the fact that many women who had undergone DXA testing did not meet the criteria for this type of screening. One study of 612 such women found that two out of every five were needlessly screened; one in every six did not meet the criteria for treatment but still were undergoing antiresorptive therapy; and one in every thirteen were receiving two antiresorptive agents. Guidelines defined by professional organizations were often ignored in clinical practice.[118] Nor did therapeutic recommendations remain static. Calcium and vitamin D supplementation had long been regarded as a vital component in the prevention and treatment of osteoporosis. In 2005 a meta-analysis of RCTs evaluating its effectiveness concluded that vitamin D supplementation of between 700 to 800 IU per day reduced the risk of hip and non-vertebral fractures in elderly ambulatory or institutionalized patients.[119] Yet seven years later, the USPSTF issued a draft statement noting that current evidence was insufficient to assess the balance of benefits and harms of daily vitamin D and calcium supplements for the primary prevention of fractures in non-institutionalized postmenopausal women.[120] In a special issue of *Calcified Tissue International* in 2013, the editors observed that there were "many unanswered questions regarding the optimal circulating levels of vitamin D that are required for bone health and for general health." Nor was there much "evidence-based information on the best way to deliver vitamin D therapeutically."[121] "The current dilemma of defining vitamin D insufficiency and identifying safe and efficacious repletion regimens," noted three other contributors, "needs to be resolved."[122]

The efforts of the osteoporosis community and the pharmaceutical industry to persuade women about the dangers of osteoporosis and the availability of treatment options, however, were far from successful. Non-adherence to the prescribed medications remained a serious problem. In a follow-up study of patients taking one of the bisphosphonates, compliance declined steadily. Among patients with zero to two years of follow-up, slightly more than half ceased taking their medication, a trend that continued over the fourteen years covered by the study.[123] Why were so many women reluctant to begin or continue pharmacotherapy? Surveys of their attitudes were revealing. Many women did not believe that treatment was necessary, the medications were safe, or the cost of drugs was reasonable. Doubts about physicians' competence and their tendency to prescribe multiple medications played a role. The fact that many women were in good health made them reluctant to go on a drug regimen. Breast cancer and thromboembolic events (a blood clot breaking loose and blocking a blood vessel) were dreaded more and seen as deadlier than osteoporosis. For those who began pharmacotherapy, many of the side effects were disheartening, to say nothing of the requirement that after taking some of the bisphosphonates, a person had to remain standing for thirty to sixty minutes. Even those women who sustained a fracture after falling felt that the fracture was due to the fall rather than to osteoporosis.[124]

The medical response to women's reluctance to undergo treatment for osteoporosis generally reflected a paternalistic attitude, as well as the belief that the legitimacy and parameters of the disease, and the efficacy of medications to treat it, had been clearly established. In the opinion of those involved in health care, too many women were unaware of the risk factors for and consequences of osteoporosis. A typical view was expressed by investigators surveying women's views of fractures in relation to bone health at midlife. "Only when there is more awareness of poor bone health as a disease process and fractures as markers for bone fragility," they wrote, "will women, men, and health providers take action to prevent future fractures and established bone disease."[125] To educate women about osteoporosis, its treatment options, the effectiveness of medications for it, and the side effects of therapy required both trust between patient and physician and shared decision making. The underlying assumption was that medical research had clearly established the legitimacy of osteoporosis, and all that remained was to ensure that knowledge about this diagnostic disease category was widely disseminated to the public.

The resistance of many women to medical authority was not the only barrier facing the osteoporosis community. There had always been a minority current of dissent that challenged many of the claims about the diagnosis of osteoporosis and the therapies for it. Women activists had often been critical of medical claims. In her 1997 book on female hormones, Susan Love noted that the definition of osteoporosis had gone through many permutations over the years. When her book was published, the most recent definition made osteoporosis a disease characterized by low bone mass and microarchitectural deterioration of bone tissue, thus leading to increased bone fragility and a consequent increase in fracture risks. To Love, the problem with this definition was that the "disease" was not an actual fracture, but only an increased risk of a fracture. It was akin to defining heart disease as having high cholesterol, rather than having a heart attack. This new definition, according to Love, simply increased the population—both women and men—who had "osteoporosis." Hip fractures, which represented the greatest danger, overwhelmingly occurred among people over the age of 70. These fractures were related not just to bone density, but to falls resulting from balance problems, dizziness, and frailty. Love was especially critical of the assumption that prevention should be based on taking a drug for many years (although she endorsed estrogen therapy). "My guess," she wrote, "is that this disease, which has been such a hot topic in recent years, will become less important as new treatments make drugs that promise a shaky sort of 'prevention' unnecessary and obsolete."[126] The National Women's Health Network echoed many of these themes. BMD testing, "either intentionally or coincidently, hooks women into a system in which drug prescriptions are their most likely choice." Nor was it a coincidence that much of the information on the dangers of osteoporosis, and on the values of bone screening and bone screening equipment, was paid for by pharmaceutical companies.[127]

Similar criticisms came from Gillian Sanson and Norton Hadler. Both questioned the very legitimacy of the diagnostic category of osteoporosis. Sanson was familiar with the medical literature dealing with osteoporosis and insisted that a massive preventive industry had emerged, based on the "myth of risk." She was critical of the assumption that women currently turning 50 would have the same fracture rates as those age 75 and above. The latter were born between 1903 and 1926, and their bone mass was developed during the Great Depression and two world wars, with this period's associated poverty, compromised nutrition, and disrupted lifestyles. Women only just entering menopause had very different lives; dietary and lifestyle changes, their control over reproduction, and a greater number of menstrual periods all potentially affected their bone health for the better. Sanson was equally critical of drug treatments for osteoporosis and the claim that they were safe.[128] Her views were echoed in the writings of Norton Hadler (a professor of medicine and microbiology/immunology at the University of North Carolina at Chapel Hill). In *Rethinking Aging*, Hadler wrote that osteoporosis was "one of the best examples of disease mongering." Bone loss, a normal attribute of aging, "has been turned into a horrifying disease that must be thwarted, regardless of the cost or the risk. . . . I have also long argued that we have no right to medicalize women, or menopause, or aging in this fashion. Medicalization redefines non-medical problems as diseases or disorders that require medical intervention when there is no reason to assert that such intervention is beneficial."[129]

One of the basic themes of critics was the fact that a screening for osteoporosis and an assessment of fracture risk was really a process of biomedicalizing aging and bone health. Applying these techniques to a new population of older women resulted in expanding the number who were labeled "at risk" and thus increasing the demand for medical tests and medication for the prevention of osteoporosis.[130] In dealing with this same issue, the *British Medical Journal* (*BMJ*) published two pieces that were critical of the trend to medicalize risk factors. Fiona Godlee, the journal's editor, pointed out that half of the world's postmenopausal women had BMD measurements that were slightly below normal. "These women are at low risk of fracture but are considered by some to be 'at risk of being at risk.'" She argued that by emphasizing relative rather than absolute risk, the benefits of treatment for these low-risk women were overstated and the harm the therapies could cause was underplayed.[131]

In the same issue of *BMJ*, Pablo Alonso-Coello and his colleagues insisted that a global alliance of drug companies, doctors, and sponsored advocacy groups promoted osteoporosis as a silent but deadly epidemic that brought misery to millions of postmenopausal women. To these authors, such a presentation represented a classic case of disease mongering: "a risk factor that has been transformed into a medical disease in order to sell tests and drugs to relatively healthy women." They were extraordinarily critical of the manner in which statistics were employed to support the importance of drug therapy for low-risk women. The authors cited an article that reanalyzed four studies of raloxifene and reported a 75 percent reduction in relative risk. In absolute terms, the reduction was a mere 0.9 percent. One of the four raloxifene studies found the greatest decrease in risk to be an 8.6 percent reduction over three to five years. This meant that up to 270 women with pre-osteoporosis might need to be treated with drugs for three years to prevent a single vertebral fracture, and most of vertebral fractures were either subclinical or asymptomatic. Moreover, virtually all of the drug studies were funded by the pharmaceutical industry, creating potential conflicts of interest.[132]

Marcia Angell (a former editor of the *New England Journal of Medicine*) noted that, beginning in the 1990s, the pharmaceutical industry "gained unprecedented control over the evaluation of its own products. . . . They often design the studies, perform the analysis, write the papers, and decide whether, when, and in what form to publish the results."[133] Studies supported by pharmaceutical firms were far more likely to produce positive findings; clinical trials that were not industry-sponsored were much less favorable.[134] Drug companies only published the favorable results; negative results—which are extremely important—went unpublished. Nor were these companies willing to release data from trials that did not have positive conclusions. Moreover, the design of many clinical trials left much to be desired. Patients recruited into such trials were usually younger, had a single diagnosis, and had fewer other health problems than individuals seen in clinical practice, all of which distorted the outcomes of the trials. Drugs were not evaluated against other drugs, but against a placebo, which not only enhanced the chances of a positive outcome, but also often gave rise to meaningless results. Trials were frequently stopped early, thus concealing adverse effects. They were also biased because dropouts, who may have experienced side effects or found that the drug did not work, were excluded from the trial, thus magnifying its positive results. Marketing techniques designed solely to influence clinical decision making rather than convey accurate information added to the perversion of good science.[135] Despite major shortcoming in the evaluation of drugs to treat osteoporosis, and the ambivalent character of the diagnoses of osteopenia and osteoporosis, the financial power and influence of the pharmaceutical industry continued to influence clinical practice in ways that left much to be desired.

FINIS

"In the final analysis," Fuller Albright wrote more than half a century ago, "very little is known about anything, and much that seems true today turns out to be only partly true tomorrow."[136] Medicine is one such example of this generalization. It is generally viewed as embodying inevitable scientific progress. The reality, however, is that failure much more accurately portrays its history. Therapies that are employed and held in high regard in one era are frequently consigned to oblivion later, because they are shown to be either ineffective or dangerous. In an analysis of a classic medical textbook published in 1927, Paul B. Beeson (a distinguished academic physician who chaired the Yale Department of Medicine and subsequently became the Nuffield Professor Medicine at Oxford University) noted that most of the recommended treatments of the pre–World War II era had disappeared by 1975. For 362 of the diseases listed in 1927, 211 therapies were found to be either harmful, useless, of questionable value, or merely symptomatic; only 23 were effective or preventive.[137] In our own time period, many therapies that were initially hailed as breakthroughs have quickly disappeared from medical practice. Other therapies, even in the absence of proof of their efficacy, nonetheless persist. The same is true of diagnostic categories. One has only to look at the *International Classification of Diseases*. The first edition was published in 1900, and the book subsequently underwent nine revisions (with a tenth now underway). Yet it has not resolved the problem of developing clear and unambiguous diagnostic categories. Kerr L. White noted that there are thousands of "labels" describing the health problems that beset humanity, but there is "no coherent conceptual or organizing theme, to say nothing of theory, and yet this classification and its modifications seek to meet the needs of policy makers, statisticians, third-party payers, managers, clinicians, and investigators of all persuasions and preoccupations in a wide range of socioeconomic and cultural settings around the world."[138]

The emphasis on prevention, however attractive, is also fraught with pitfalls. As David L. Sackett (one of the most distinguished clinical epidemiologists and a pioneer in evidence-based medicine) has observed, preventive medicine displays three elements of arrogance. First, it is aggressively assertive and pursues symptomless individuals, telling them what they must do to remain healthy. Second, it is presumptuous in its confident belief that the interventions it espouses will do more good than harm. Finally, it is overbearing in attacking those who question the value of its recommendations. Sackett argued that preventive medicine is fundamentally different from curative medicine, which seeks to treat symptomatic patients without guaranteeing the success of the intervention.[139]

The history of the diagnostic category of osteoporosis, and of the therapies developed to prevent and treat it, embodies these themes and illustrates many of the problems and dilemmas of modern medicine. An increase in the proportion of the elderly in the population heightened sensitivity toward their health needs and problems. The diagnosis of osteoporosis, in its modern form, originated when the nascent specialty of endocrinology shed light on menopause and stimulated efforts to illuminate bone physiology and modeling. During that period, menopause was viewed less as a normal stage of life and more as a condition of estrogen deficiency, and osteoporosis was a relatively narrow diagnosis that, at first, was confined to a small subset of postmenopausal women with severe non-traumatic vertebral fractures that led to disability. Figures like Fuller Albright and others were primarily concerned with women who developed severe fractures that in many instances incapacitated them, but, extrapolating from that, they did consider osteoporosis to be a major problem in the broader population.

During and after the 1980s, the diagnostic category of osteoporosis was invested with new meanings. A culture of youth, a belief that many of the effects of aging could be postponed, faith in medical authority, the transference of funding for research and clinical trials from more impartial sources to drug companies, and mass marketing by the pharmaceutical industry—first of HRT and subsequently of a variety of antiosteoporotic drugs— all elevated the significance of osteoporosis and dramatically expanded its boundaries. Initially the focus was on vertebral fractures, but soon hip fractures took center stage. At the same time, the introduction of new scanning technologies made it possible to view bones in novel ways. Slowly but surely, BMD testing became the primary means of diagnosing osteoporosis. Although the WHO study group that created the T-score in the 1990s conceded that its cutoff values were "somewhat arbitrary," the scale was quickly adopted, because it allegedly provided a firm foundation for the screening and treatment of osteoporosis.[140] T-score values sufficient to prompt screening and treatment, however, were continuously modified. The end result was a recommendation that virtually all older persons be screened, and that a very large proportion of this group should undergo pharmacological treatment. The methodological weaknesses of many epidemiological studies, the fact that evidence linking low BMD to high fracture rates was hardly persuasive, and the relatively small benefits of medications in the prevention and treatment of osteoporosis were largely overlooked or ignored altogether. The public was besieged by messages warning of the dangers posed by the osteoporosis "epidemic" and emphasizing the importance of screening and pharmacotherapy to combat this disease. Disagreements within the osteoporosis community about the values of screening and of antiosteoporotic drugs remained largely hidden from the lay public. Clinicians in a variety of specialties remained more or less oblivious to the indeterminate state of knowledge about bone physiology and bone development; they took their cue from presumed experts and pharmaceutical representatives.

Thus by the beginning of the twenty-first century, medicine's original concern with a small group of women with a history of non-traumatic fractures and subsequent disabilities had been lost from view. Osteopenia and osteoporosis now included virtually all postmenopausal women, plus many men in the same age category. The risk of osteoporosis and the disease itself became conflated, and dire warnings about the dangers of osteoporosis increased in intensity. Tens of millions of asymptomatic persons were urged to undergo screening and take medications that would presumably prevent the disease. Whether a broadening of the boundaries of the diagnostic category of osteopenia and osteoporosis will result in better outcomes remains to be determined. Similarly, only the future can tell whether the legitimacy of the diagnosis of osteoporosis will persist.

NOTES

Chapter 1. History and Demography

1. See Bernard Straus, "Problems and Progress in Geriatrics," *Journal of the American Geriatrics Society*, 16 (1968): 257–66.

2. William J. Rolfe, ed., *Shakespeare's Sonnets* (New York: Harper & Brothers, 1884), 83.

3. Prov. 17:22; Job 21:24; Eccles. 11:5.

4. Plato, *Timaeus*, ed. and trans. John Warrington (London: J. M. Dent, 1965), 105.

5. My discussion of historical concepts of bone formation is drawn from Donald H. Enlow, *Principles of Bone Remodeling* (Springfield, IL: Charles C Thomas, 1963), 3–22; and Henry M. Frost, "Osteoporosis: Quo Vadis?," *Orthopedic Clinics of North America*, 12 (1981): 683–91.

6. D. Schapira and C. Schapira, "Osteoporosis: The Evolution of a Scientific Term," *Osteoporosis International*, 2 (1992): 164–67.

7. Astley Cooper, *A Treatise on Dislocations and Fractures of Joints*, 2nd American ed. (from the 6th London ed.) (Boston: Lilly & Wait, 1832; original ed. 1822), 129.

8. Dr. Bennett, "Senile Osteoporosis," *Dublin Journal of Medical Science*, 3rd ser., 66 (1878): 272–74.

9. Dr. Wilks, "Case of Osteoporosis, or Spongy Hypertrophy of the Bones," *Transactions of the Pathological Society of London*, 20 (1869): 273–77.

10. René Fontaine and Louis G. Hermann, "Post-Traumatic Osteoporosis," *Annals of Surgery*, 97 (1933): 26–61. Gustav Pommer made this distinction in his *Untersuchengen über Osteomalacie und Rachitis* (Leipzig, Germany: Vogel, 1885).

11. David H. Fischer, *Growing Old in America*, expanded ed. (New York: Oxford University Press, 1978), 28, 272.

12. Ibid., 29–53; Carole Haber, *Beyond Sixty-Five: The Dilemma of Old Age in America* (Cambridge: Cambridge University Press, 1983), 15–18. See also Philip Greven, *Four Generations: Population, Land, and Family in Colonial Andover, Massachusetts* (Ithaca, NY: Cornell University Press, 1970).

13. Jefferson to Adams, 1 August 1816, 7 October 1818, 15 August 1820, 12 October 1823, and Adams to Jefferson, 11 June 1822, 14 January 1826, in *The Adams-Jefferson Letters*, ed. Lester J. Cappon, 2 vols. (Chapel Hill: University of North Carolina Press, 1959), vol. 2: 483–84, 528, 565, 579, 599, 613.

14. Haber, *Beyond Sixty-Five*, 28–34.

15. Stephen Katz, *Disciplining Old Age: The Formation of Gerontological Knowledge* (Charlottesville: University Press of Virginia, 1996), 41.

16. See Theodore M. Porter, *The Rise of Statistical Thinking, 1820–1900* (Princeton, NJ: Princeton University Press, 1986).

17. Benjamin Gompertz, "On the Nature of the Function Expressive of the Law of Human Mortality, and on a New Mode of Determining the Value of Life Contingencies," *Philosophical Transactions of the Royal Society of London*, 115 (1825): 513–83. For a discussion, see S. Jay Olshansky and Bruce A. Carnes, "Ever since Gompertz," *Demography*, 34 (1997): 1–15.

18. Adolphe Quetelet, *Research on the Propensity for Crime at Different Ages*, trans. Sawyer F. Sylvester (Cincinnati: Anderson, 1984), 54–55.

19. Adolphe Quetelet, *A Treatise on Man and the Development of his Faculties*, trans. under the supervision of R. Knox (Edinburgh: William and Robert Chambers, 1842), 5, 75.

20. Haber, *Beyond Sixty-Five*, 45–63.

21. J. S. Nowlin, "The Climacteric: Its Phenomena and Dangers," *Nashville Journal of Medicine and Surgery*, 77 (1895): 7–13; George L. Sinclair, "Climacteric Insanity," *International Clinics*, 3rd ser., 4 (1894): 147–52; and George H. Rohé, "The Mental Disturbances of the Climacteric Period," *Maryland Medical Journal*, 34 (1895): 258–61.

22. George H. Candler, "The Recognition and Treatment of 'Climacteric Disorder' in the Male," *American Journal of Clinical Medicine*, 26 (1919): 694–700; and John S. Turner, "The Male Climacteric," *Texas State Medical Journal*, 12 (1916), 251–54.

23. Ignatz L. Nascher, "The Senile Climacteric," *New York Medical Journal*, 94 (1911): 1125–26; "The Male Climacteric," *American Medicine*, 27 (1942): 242–47; and "Geriatrics," *New York Medical Journal*, 90 (1909): 358–59.

24. Andrew Achenbaum, *Old Age in the New Land: The American Experience since 1790* (Baltimore: Johns Hopkins University Press, 1978), 44–45.

25. Carl Snyder, "The Quest of Prolonged Youth," *Living Age*, 251 (1906): 323–37.

26. Fischer, *Growing Old*, 113–56.

27. Ibid., 113.

28. George M. Beard, *Legal Responsibility in Old Age, Based on Researches into the Relation of Age to Work* (New York: Russells', 1874), 7–12; and Beard, *American Nervousness: Its Causes and Consequences* (New York: G. P. Putnam's, 1881), 249–52.

29. William Osler, "The Fixed Period," *Aequanimitas* (Philadelphia, P. Blakiston's Son, 1906), reprinted in *The "Fixed Period" Controversy*, ed. Gerald J. Gruman (New York: Arno Press, 1970), 381–83. See esp. the discussion in Thomas R. Cole, *The Journey of Life: A Cultural History of Aging in America* (Cambridge: Cambridge University Press, 1992), 161–90.

30. Snyder, "Quest of Prolonged Youth," 323; and Carl Snyder, *New Conceptions in Science* (New York: Harper & Brothers, 1903), 331–55.

31. G. Stanley Hall, "Old Age," *Atlantic Monthly*, 127 (1921): 23–31.

32. G. Stanley Hall, *Senescence: The Last Half of Life* (New York: D. Appleton, 1923), 403, 419.

33. These themes are dealt with in Haber, *Beyond Sixty-Five*, and Achenbaum, *Old Age*.

34. Achenbaum, *Old Age*, 47–51.

35. Franklin MacVeagh, "Civil Service Pensions," *Annals of the American Academy of Political and Social Science*, 38 (1911): 3–5.

36. F. Spencer Baldwin, "Retirement Systems for Municipal Employees," *Annals of the American Academy of Political and Social Science*, 38 (1911): 6–14. See also William Graebner, *A History of Retirement: The Meaning and Function of an American Institution, 1885–1978* (New Haven, CT: Yale University Press, 1980).

37. U.S. Bureau of the Census, *Statistical Abstract of the United States, 1996* (Washington, DC: U.S. Government Printing Office, 1996), 14–16.

38. National Center for Health Statistics, *Health, United States, 2000* (Washington, DC: U.S. Government Printing Office, 2000), 160; and U.S. Bureau of the Census, *Statistical Abstract of the United States, 1998* (Washington, DC: U.S. Government Printing Office, 1998), 16–17.

39. Carl L. Erhardt and Joyce E. Berlin, eds., *Mortality and Morbidity in the United States* (Cambridge, MA: Harvard University Press, 1974), 4–5.

40. Gerald N. Grob, *The Deadly Truth: A History of Disease in America* (Cambridge, MA: Harvard University Press, 2002), 180–216.

41. Sherry L. Murphy, "Deaths: Final Data for 1998," *National Vital Statistics Reports*, 48, no. 11 (Jul. 24, 2000): 18.

42. George H. Bigelow and Herbert L. Lombard, *Cancer and Other Chronic Diseases in Massachusetts* (Boston: Houghton Mifflin, 1933), 1–4, 10.

43. Louis I. Dublin and Alfred J. Lotka, *Twenty-Five Years of Health Progress: A Study of the Mortality Experience among the Industrial Policy-Holders of the Metropolitan Life Insurance Company, 1911 to 1935* (New York: Metropolitan Life Insurance, 1937), 4, 119–42; and Ernst P. Boas, *The Unseen Plague: Chronic Disease* (New York: J. J. Augustin, 1940), 4, 17–18, 121.

44. Joseph C. Bloodgood, "Bone Diseases, Osteoporosis or Lipomasia from Fixation and Non-Use," *Journal of Radiology*, 3 (1922): 403–6; Duval Prey and John M. Foster, "Post-Traumatic Osteoporosis of the Carpal Bones," *Colorado Medicine*, 31 (1934): 86–90; and Fontaine and Hermann, "Post-Traumatic Painful Osteoporosis," 26–61.

45. William Osler, *The Principles and Practice of Medicine*, 1st ed. (New York: D. Appleton, 1892); 4th ed. (1901); 6th ed. (1906); Henry A. Christian, *The Principles and Practice of Medicine*, 15th ed. (New York: Appleton-Century, 1944), 1202–8; and Logan Clendening, *Modern Methods of Treatment* (Saint Louis: C. V. Mosby, 1924).

Chapter 2. The Origins of a Diagnosis

1. John Howship, "Observations on the Morbid Structure of Bones, and Attempt at an Arrangement of Their Diseases," *Transactions of the Medical and Chirurgical Society of London*, 8 (1817): 57–107.

2. J. Albert Key, "Bone Atrophy and Absorption," *International Journal of Orthodontics*, 15 (1929): 949–82.

3. Chandak Sengoopta, "The Modern Ovary: Constructions, Meanings, Uses," *History of Science*, 38 (2000): 426–29. See also Carroll Smith-Rosenberg and Charles Rosenberg, "The Female Animal: Medical and Biological Views of Woman and Her Role in Nineteenth-Century America," *Journal of American History*, 60 (1973): 332–56.

4. This discussion is based on Ludmilla Jordanova, *Sexual Visions: Images of Gender in Science and Medicine between the Eighteenth and Twentieth Centuries* (Madison: University of Wisconsin Press, 1989); and Margaret Lock, *Encounters with Aging: Mythologies of Menopause in Japan and North America* (Berkeley: University of California Press, 1993).

5. Merriley Borell, "Brown-Séquard's Organotherapy and Its Appearance in America at the End of the Nineteenth Century," *Bulletin of the History of Medicine*, 50 (1976), 309–20.

6. George W. Corner, "The Early History of the Oestrogenic Hormones," *Journal of Endocrinology*, 31 (1965): iii–vi.

7. For an understanding of the evolution of endocrinology, I have relied on Elizabeth S. Watkins, *The Estrogen Elixir: A History of Hormone Replacement Therapy in America* (Baltimore: Johns Hopkins University Press, 2007); S. M. McCann, ed., *Endocrinology: People and Ideas* (Bethesda, MD: American Physiological Society, 1988); and Victor C. Medvei, *The History of Clinical Endocrinology*, 2nd ed., rev. (Carnforth, UK: Parthenon, 1993).

8. Harvey Cushing, "Disorders of the Pituitary Gland: Retrospective and Prophetic," *Journal of the American Medical Association*, 76 (1921): 1721–26; and Hans Lisser, "The Endocrine Society: The First Forty Years (1917–1957)," *Endocrinology*, 80 (1967): 5–28.

9. Watkins, *Estrogen Elixir*, 10–20.

10. Ibid.

11. Robert T. Frank, "Treatment of Disorders of the Menopause," *Bulletin of the New York Academy of Medicine*, 17 (1941): 854–63. For Frank's career in endocrinology, see Corner, "Early History," x–xi.

12. Fuller Albright and Read Ellsworth, "Studies on the Physiology of the Parathyroid Glands: Calcium and Phosphorous Studies on a Case of Idiopathic Hypoparathyroidism," *Journal of Clinical Investigation*, 7 (1929): 183–201; and Theodore H. Ingalls, Gordon Donaldson, and Fuller Albright, "The Locus of Action of the Parathyroid Hormone: Experimental Studies with Parathyroid Extract on Normal and Nephrectomized Rats," *Journal of Clinical Investigation*, 22 (1943): 603–8. In 1929 Albright, together with Ellsworth, Joseph C. Aub, Walter Bauer, Clark Health, and Marion Ropes, published five articles (one in the *Journal of Experimental Medicine* and the others in the *Journal of Clinical Investigation*) dealing largely with calcium and phosphorus metabolism.

13. Oscar Riddle and Warren H. Reinhart, "Studies on the Physiology of Reproduction in Birds: Blood Calcium Changes in the Reproductive Cycle," *American Journal of Physiology*, 76 (1926): 660–76.

14. Preston Keys and Truman S. Potter, "Physiologic Marrow Ossification in Female Pigeons," *Anatomical Record*, 60 (1934): 377–79.

15. Oscar Riddle and Louis B. Dotti, "Blood Calcium in Relation to Anterior Pituitary and Sex Hormones," *Science*, 84 (1936): 557–59.

16. Carroll A. Pfeiffer and William U. Gardner, "Skeletal Changes and Blood Serum Calcium Level in Pigeons Receiving Estrogens," *Endocrinology*, 23 (1938): 485–91; Gardner and Pfeiffer, "Skeletal Changes in Mice Receiving Estrogen," *Proceedings of the Society for Experimental Biology and Medicine*, 37 (1938): 678–79; and Gardner and Pfeiffer, "Influence of Estrogens and Androgens on the Skeletal System," *Physiological Reviews*, 23 (1943): 139–65.

17. See, for example, Max Ballin, "Skeletal Pathology of Endocrine Origins," *Annals of Surgery*, 98 (1933): 868–81.

18. Henry A. Harris, *Bone Growth in Health and Disease* (London: Oxford University Press, 1933), 227. For a discussion of the problems in explaining the growth of bones, see James C. Brash, "Some Problems in the Growth and Developmental Mechanics of Bone," *Edinburgh Medical Journal*, 41 (1934): 305–18, 363–87.

19. Ernst Lachman and Mary Whelan, "The Roentgen Diagnosis of Osteoporosis and Its Limitations," *Radiology*, 26 (1936): 165–77.

20. Fuller Albright, "Some of the 'Do's' and 'Do-Nots' in Clinical Investigation," *Journal of Clinical Investigation*, 23 (1944): 921–26. My description of Albright's career is drawn from A. Leaf, "Fuller Albright, 1900–1969," *Biographical Memoirs of the National Academy of Sciences*, 48 (1976): 1–22; and Theodore B. Schwartz, "How to Learn from Patients: Fuller Albright's Exploration of Adrenal Function," *Annals of Internal Medicine*, 123 (1995): 225–29.

21. Fuller Albright, Esther Bloomberg, and Patricia H. Smith, "Postmenopausal Osteoporosis," *Transactions of the Association of American Physicians*, 55 (1940): 298–305; and Albright, Smith, and Anna M. Richardson, "Postmenopausal Osteoporosis: Its Clinical Features," *Journal of the American Medical Association*, 114 (1941): 2465–74.

22. For discussions of Albright's work, see Gilbert S. Gordon, "Fuller Albright and Postmenopausal Osteoporosis: A Personal Appreciation," *Perspectives in Biology and Medicine*, 24 (1981): 547–60; and Anne P. Forbes, "Fuller Albright: His Concept of Postmenopausal Osteoporosis and What Came of It," *Clinical Orthopaedics and Related Research*, 269 (1991): 128–41.

23. Albright, Smith, and Richardson, "Postmenopausal Osteoporosis," 2472; and Fuller Albright, "Osteoporosis," *Annals of Internal Medicine*, 27 (1947): 865.

24. Edward C. Reifenstein Jr. and Fuller Albright, "The Metabolic Effects of Steroid Hormones in Osteoporosis," *Journal of Clinical Investigation*, 26 (1947): 24–56; and Albright and Reifenstein, *The Parathyroid Glands and Metabolic Bone Disease* (Baltimore: Williams & Wilkins: 1948), 145–204.

25. Philip H. Henneman and Stanley Wallach, "A Review of the Prolonged Use of Estrogens and Androgens in Postmenopausal and Senile Osteoporosis," *AMA Archives of Internal Medicine*, 100 (1957): 715–23.

26. Ian A. Anderson, "Postmenopausal Osteoporosis, Clinical Manifestations, and the Treatment with Oestrogens," *Quarterly Journal of Medicine*, n.s., 18 (1949): 67–96.

27. Albright, "Osteoporosis," 882.

28. See Harold M. Frost, "The Evolution of Pathophysiologic Knowledge of Osteoporoses," *Orthopedic Clinics of North America*, 12 (1981): 473–81.

29. Franklin C. Mclean, "Calcified Tissue Research: 1967," *Calcified Tissue Research*, 1 (1967): 1–7; and Felix Bronner, "Thirty-One Years of Bones and Teeth— Retrospect and Prospect," *Calcified Tissue International*, 37 (1985): 3–5.

30. R. H. Follis, "Osteoporosis," *Bulletin on Rheumatic Diseases*, 4 (1954): 69–70.

31. Richmond W. Smith Jr., William R. Eyler, and Raymond C. Mellinger, "On the Incidence of Senile Osteoporosis," *Annals of Internal Medicine*, 52 (1960): 773–81; Marc Moldawer, Stanley J. Zimmerman, and Lois C. Collins, "Incidence of Osteoporosis in Elderly Whites and Elderly Negroes," *JAMA*, 194 (1965): 859–62; Alfred J. Bollet, Gerald Engh, and William Parson, "Epidemiology of Osteoporosis," *Archives of Internal Medicine*, 116 (1965): 191–94; and Smith and Juan Rizek, "Epidemiologic Studies of Osteoporosis in Women of Puerto Rico and Southeastern Michigan with Special Reference to Age, Race, National Origin, and to Other Related or Associated Findings," *Clinical Orthopaedics and Related Research*, 45 (1966): 31–48.

32. B. E. C. Nordin, "International Patterns of Osteoporosis," *Clinical Orthopaedics and Related Research*, 45 (1966): 17–30.

33. G. Donald Whedon, "International Collaborative Study of Osteoporosis and Fracture Epidemiology," *Clinical Orthopaedics and Related Research*, 45 (1966): 13–15. A search of the PubMed biomedical literature database (www.ncbi.nlm.nih .gov/pubmed/) covering the period 1966–1972 could find no references to a WHO study of osteoporosis and fracture epidemiology.

34. Edward C. Reifenstein Jr., "The Relationships of Steroid Hormones to the Development and the Management of Osteoporosis in Aging People," *Clinical Orthopaedics*, 10 (1957): 206–53; Reifenstein, "Steroid Therapy for Osteoporosis," *Geriatrics*, 12 (1957): 139–40; and Reifenstein, "Control of Corticoid-Induced Protein Depletion and Osteoporosis by Anabolic Steroid Therapy," *Metabolism*, 7 (1958): 78–89. For a critical analysis of Albright's theory, see Robert P. Heaney, "A Unified Concept of Osteoporosis," *American Journal of Medicine*, 39 (1965): 877–80.

35. Marshall R. Urist, "The Etiology of Osteoporosis," *Journal of the American Medical Association*, 169 (1959): 710–12; Urist, "Observations Bearing on the Problem of Osteoporosis," in *Bone as a Tissue*, ed. Kaare Rodahl, Jesse T. Nicholson, and Ernest M. Brown (New York: McGraw Hill, 1960), 42–43; B. E. C. Nordin, "The Pathogenesis of Osteoporosis," *Lancet*, 277 (May 13, 1961): 1011–15; Nordin,

J. MacGregor, and D. A. Smith, "The Incidence of Osteoporosis in Normal Women: Its Relation to Age and the Menopause," *Quarterly Journal of Medicine*, 35 (1966): 25–38; and "Calcium Metabolism in Osteoporosis," *Nutrition Reviews*, 19 (1961): 269–72.

36. For a sampling of differing explanations of bone development and the etiology of osteoporosis in the 1950s, see Esther M. Greisheimer, "Postmenopausal Osteoporosis," *Journal of the American Medical Women's Association*, 6 (1951): 183–84; L. Snapper, "Osteoporosis," *Medical Clinics of North America*, 36 (1952): 847–63; Frederic C. Bartter, "Osteoporosis," *American Journal of Medicine*, 22 (1957): 797–806; Edmund R. Yendt, "Osteoporosis," *Postgraduate Medicine*, 22 (1957): 38–45; "Panel Discussion on Osteoporosis," *Journal of the American Geriatrics Society*, 5 (1957): 363–84; and Charles H. Epps, "Current Concepts in Osteoporosis," *Journal of the National Medical Association*, 51 (1959): 106–12.

37. B. E. F. Nordin, "The Application of Basic Science to Osteoporosis," in *Bone Biodynamics*, ed. Harold M. Frost (Boston: Little, Brown, 1964), 521–42.

38. Franklin McLean and Marshall R. Urist, *Bone: Fundamentals of the Physiology of Skeletal Tissue*, 3rd ed. (Chicago: University of Chicago Press, 1966), 239–40; and Urist, "Rarefying Disease of the Skeleton: Observations Dealing with Aged and Dead Bone in Patients with Osteoporosis," in *Mechanisms of Hard Tissue Destruction*, ed. Reidar E. Sognnaes (Washington, DC: American Association for the Advancement of Science, 1963), 386–446.

39. G. Donald Whedon, "Symposium Comment," in *Osteoporosis*, ed. Uriel S. Barsel (New York: Grune & Stratton, 1970), 266.

40. H. F. Newton Jones and D. B. Morgan, "Osteoporosis: Disease or Senescence?," *Lancet*, 291 (Feb. 3, 1968): 232–33.

41. Paul D. Saville, "Osteoporosis: Disease or Senescence?," *Lancet*, 291 (Mar. 9, 1968): 535.

42. Göran C. H. Bauer, "Kinetics of Calcium and Strontium Metabolism in Man," in *Bone as a Tissue*, ed. Rodahl, Nicholson, and Brown, 122; and Bauer, Arvid Carlsson, and Bertil Lindquist, "Metabolism and Homeostatic Function of Bone," in *Mineral Metabolism: An Advanced Treatise*, ed. C. L. Comer and Felix Bronner, 2 vols. (New York: Academic Press, 1961), vol. 1, part B: 618.

43. Nordin, MacGregor, and Smith, "Incidence of Osteoporosis," 25; and B. E. F. Nordin, "The Nature of Osteoporosis," *Transactions of the Medical Society of London*, 83 (1967): 172–79.

44. Reifenstein, "Relation of Steroid Hormones," 209–11.

45. Ibid., 211.

46. G. Alan Rose, "The Radiological Diagnosis of Osteoporosis, Osteomalacia, and Hyperparathyroidism," *Clinical Radiology*, 15 (1964): 75.

47. H. M. Hodrinson, A. N. Exton-Smith, and M. F. Crowley, "Diagnosis and Assessment of Osteoporosis," *Postgraduate Medical Journal*, 39 (1963): 433.

48. Willard E. Dotter and Lewis M. Hurxthal, "An Approach to the Diagnosis of Metabolic Bone Disease in the Elderly Patient," *Geriatrics*, 20 (1965): 424.

49. Ernest Lachman, "The Normal Radiology of Old Age and Its Borderlines," *Journal of the Oklahoma State Medical Association*, 53 (1960): 87–92; and Lachman, "Osteoporosis: The Potentialities and Limitations of Its Roentgenologic Diagnoses," *American Journal of Roentgenology, Radium Therapy, and Nuclear Medicine*, 74 (1955): 712–15.

50. McLean, "Calcified Tissue Research," 3.

51. Joseph P. Weinmann and Harry Sicher, *Bone and Bones: Fundamentals of Bone Biology* (Saint Louis: C. V. Mosby, 1947), 303–4; and 2nd ed., 1955, 328–29.

52. Frederic C. McLean and Marshall R. Urist, *Bone: An Introduction to the Physiology of Skeletal Tissue*, 2nd ed. (Chicago: University of Chicago Press, 1961), 45, 57, 220. Similar statements were repeated seven years later in McLean and Urist, *Bone: Fundamentals of the Physiology*, 56–71, 104.

53. McLean, "Calcified Tissue Research," 6.

54. Harold M. Frost, "The Osteoporoses: A Definition of Terms and Concepts," *Henry Ford Hospital Medical Bulletin*, 10 (1962): 315.

55. Ernst Aegerter and John A. Kirkpatrick, *Orthopedic Diseases: Physiology—Pathology—Radiology*, 4th ed. (Philadelphia: W. B. Saunders, 1975), 331.

56. Frost, "Evolution of Pathophysiologic Knowledge," 479–80.

57. Frost's works in the 1960s and 1970s included *Bone Remodeling Dynamics* (Springfield, IL: Charles C. Thomas, 1963); *The Laws of Bone Structure* (Springfield, IL: Charles C. Thomas, 1964); *The Bone Dynamics in Osteoporosis and Osteomalacia* (Springfield, IL: Charles C. Thomas, 1966); and *Bone Remodeling and Its Relationship to Metabolic Bone Diseases* (Springfield, IL: Charles C. Thomas, 1973). The seminal article—Robert Hattner, Bruce N. Epker, and Harold M. Frost, "Suggested Sequential Mode of Control of Changes in Cell Behaviour in Adult Bone Remodelling," *Nature*, 206 (1965): 489–90—was reprinted as a Landmark piece (with an interview) in the *Journal of NIH Research*, 7 (1995): 54–59. For biographical data and appreciations by colleagues, see "Harold M. Frost, M.D.: His Contributions," *Journal of Musculoskeletal and Neuronal Interactions*, 4 (2004): 348–56; and W. S. S. Jee, "Harold M. Frost, M.D., D.Sc. (Hon)—One Man's Association," *Journal of Musculoskeletal and Neuronal Interactions*, 6 (2006): 113–21.

58. H. E. Meema, M. L. Bunker, and Silvia Meema, "Loss of Compact Bone Due to Menopause," *Obstetrics and Gynecology*, 26 (1965): 341.

59. B. Lawrence Riggs, Jenifer Jowsey, Patrick J. Kelly, James D. Jones, and Frank T. Maher, "Effect of Sex Hormones on Bone in Primary Osteoporosis," *Journal of Clinical Investigation*, 48 (1969): 1065–72.

60. "Calcium Metabolism in Osteoporosis," 269–72; Leo Lutwak, "Osteoporosis: A Disorder of Nutrition," *New York State Journal of Medicine*, 63 (1963): 590–93; Lutwak, "Osteoporosis—a Mineral Deficiency Disease," *Journal of the American Dietetic Association*, 44 (1964): 173–75; J. Gershon-Cohen and Jenifer Jowsey, "Relationship of Dietary Calcium to Osteoporosis," *Metabolism*, 13 (1964): 221–26; Herta Spencer, J. Menczel, I. Lewin, and J. Samachson, "Absorption of Calcium

in Osteoporosis," *American Journal of Medicine*, 37 (1964): 223–34; Ole Munck, "Osteoporosis Due to Malabsorption of Calcium Responding Favourably to Large Doses of Vitamin D," *Quarterly Journal of Medicine*, 33 (1964): 209–21; and Jowsey, R. C. Hoy, C. Y. Pak, and F. C. Bartter, "The Treatment of Osteoporosis with Calcium Infusions: Evaluation of Bone Biopsies," *American Journal of Medicine*, 47 (1969): 17–22.

61. James W. Hall and B. J. Kennedy, "Idiopathic Osteoporosis," *Annals of Internal Medicine*, 108 (1961): 448–55.

62. For a sampling of the diverse therapeutic recommendations during the 1960s, see Leo Lutwak, Jacob Bassin, Boy Frame, and Harold Frost, "Osteoporosis: A Review of Pathogenesis and Treatment," *Annals of Internal Medicine*, 58 (1963): 539–50; Daniel S. Bernstein and Charles D. Guri, "Osteoporosis: Etiology and Therapy," *Postgraduate Medicine*, 34 (1963): 410–16; and Heinrich C. Haas, "Osteoporosis," *Geriatrics*, 22 (1967): 100–111.

63. Michael T. Harrison, "The Riddle of Osteoporosis [editorial]," *Journal of Chronic Diseases*, 16 (1963): 193.

64. Paul R. Lipscomb, "Fractures in the Aged," *Postgraduate Medicine*, 34 (1963): 410–11.

65. Manual Rodstein, "The Aging Process and Disease," *Nursing Outlook*, 12 (1964): 43–46.

66. Richmond W. Smith Jr. and Richard R. Walker, "Femoral Expansion in Aging Women: Implications for Osteoporosis and Fractures," *Science*, 145 (1964): 156.

67. R. A. Caldwell, "Observations on the Incidence, Aetiology, and Pathology of Senile Osteoporosis," *Journal of Clinical Pathology*, 15 (1962): 421–31; Leo Lutwak and G. Donald Whedon, "Osteoporosis," *Disease-a-Month: DM*, Apr. 1963, 5–6; M. Edward Davis, Nels M. Strandjord, and Lawrence H. Lanzl, "Estrogens and the Aging Process: The Detection, Prevention, and Retardation of Osteoporosis," *JAMA*, 196 (1966): 129; and A. W. Dunn, "Senile Osteoporosis," *Geriatrics*, 22 (1967): 175–76.

68. Luke S. W. Chu and David I. Abramson, "Diagnosis and Treatment of Osteoporosis," *Geriatrics*, 18 (1963): 681; and M. L. Riccitelli, "The Management of Osteoporosis in the Aged and Infirm," *Journal of the American Geriatrics Society*, 10 (1962): 498.

69. For a discussion of this point, see Kristin K. Barker, *The Fibromyalgia Story: Medical Authority and Women's World of Pain* (Philadelphia: Temple University Press, 2005); and Gerald N. Grob, "The Rise of Fibromyalgia in 20th-Century America," *Perspectives in Biology and Medicine*, 54 (2011): 417–37.

Chapter 3. The Transformation of Osteoporosis

1. President's Scientific Research Board, *Science and Public Policy*, 5 vols. (Washington, DC: U.S. Government Printing Office, 1947), vol. 1: 3, 115–18.

2. Robert N. Butler, "Age-Ism: Another Form of Bigotry," *Gerontologist*, 9

(1969): 243. See also Butler, *Why Survive?: Being Old in America* (New York: Harper & Row, 1975); and Butler, *The Longevity Revolution: The Benefits and Challenges of Living a Long Life* (New York: Public Affairs, 2008).

3. See esp. Sheila M. Rothman and David J. Rothman, *The Pursuit of Perfection: The Promise and Perils of Medical Enhancement* (New York: Pantheon Books, 2003); and Peter Conrad, *The Medicalization of Society: On the Transformation of Human Conditions into Treatable Medical Disorders* (Baltimore: Johns Hopkins University Press, 2007).

4. See Patricia A. Kaufert and Penny Gilbert, "Women, Menopause, and Medicalization," *Culture, Medicine and Psychiatry*, 10 (1986): 7–31; Margaret Lock, *Encounters with Aging: Mythologies of Menopause in Japan and North America* (Berkeley: University of California Press, 1993); and Elizabeth S. Watkins, *The Estrogen Elixir: A History of Hormone Replacement Therapy in America* (Baltimore: Johns Hopkins University Press, 2007).

5. Ray Moynihan, Iona Heath, and David Henry, "Selling Sickness: The Pharmaceutical Industry and Disease Mongering," *British Medical Journal*, 324 (2002): 886–91; Marcia Angell, "Industry-Sponsored Clinical Research: A Broken System," *JAMA*, 300 (2008): 1069–71; Joseph S. Ross, Kevin P. Hill, David S. Egilman, and Harlan M. Krumholz, "Guest Authorship and Ghostwriting in Publications Related to Rofecoxib: A Case Study of Industry Documents from Rofecoxib Litigation," *JAMA*, 299 (2008): 1800–1812; Catherine D. DeAngelis and Phil B. Fontanarosa, "Impugning the Integrity of Medical Science: The Adverse Effects of Industry Influence," *JAMA*, 299 (2008): 1833–35; and Marcia Angell, *The Truth about the Drug Companies: How They Deceive Us and What to Do about It* (New York: Random House, 2004).

6. Urial S. Barzel, ed., *Osteoporosis* (New York: Grune & Stratton, 1970), 123–24.

7. Ibid., 23.

8. G. Hazan, I. Leichter, A. Loewinger, A. Weinreb, and G. C. Robin, "The Early Detection of Osteoporosis by Compton Gamma Ray Spectroscopy," *Physics in Medicine and Biology*, 22 (1977): 1073–84; Dennis D. Patton, "Radionuclide Bone Scanning in Diseases of the Spine," *Radiologic Clinics of North America*, 15 (1977): 177–201; Helen W. Wahner, B. Lawrence Riggs, and John W. Beabout, "Diagnosis of Osteoporosis: Usefulness of Photon Absorptiometry of the Radius," *Journal of Nuclear Medicine*, 18 (1977): 432–37; R. Lindsay and J. B. Anderson, "Radiological Determination of Changes in Bone Mineral Content," *Radiography*, 44 (1978): 21–26; and Marshall A. Weissberger, "Computed Tomography Scanning for the Measurement of Bone Mineral in the Human Spine," *Journal of Computer Assisted Tomography*, 2 (1978): 253–62. For a discussion of the technological imaging techniques of the 1970s, see Gilbert S. Gordan and Cynthia Vaughan, *Clinical Management of the Osteoporoses* (Acton, MA: Publishing Sciences Group, 1976), 64–73.

9. Watkins, *Estrogen Elixir*, 172.

10. W. Tabb Moore, "The Evaluation of Bone Density Findings in Normal Pop-

ulations and Osteoporosis," *Transactions of the American Clinical and Climatological Association*, 86 (1975): 128–38; L. Nilas, J. Borg, A. Gotfredsen, and C. Christiansen, "Comparison of Single- and Dual-Photon Absorptiometry in Postmenopausal Bone Mineral Loss," *Journal of Nuclear Medicine*, 26 (1985): 1257–62; and Stanley M. Garn, Andrew K. Poznanski, and Jerrold M. Nagy, "Bone Measurement in the Differential Diagnosis of Osteopenia and Osteoporosis," *Radiology*, 100 (1971): 509–18.

11. Anthony A. Albanese, A. H. Edelson, E. J. Lorenze Jr., M. L. Woodhull, and E. H. Wein, "Problems of Bone Health in Elderly: Ten-Year Study," *New York State Journal of Medicine*, 75 (1976): 326–36.

12. Friedrich Kuhlencordt, "Osteoporosis—a Clinical Review," *Calcified Tissue Research*, 21, Suppl. (1976): 405–6.

13. Peter Adams, G. T. Davies, and F. Sweetnam, "Osteoporosis and the Effects of Aging on Bone Mass in Elderly Men and Women," *Quarterly Journal of Medicine*, n.s., 39 (1970): 614.

14. Joseph P. Whalen, "Bone Mineral Management [editorial]," *American Journal of Roentgenology*, 126 (1976): 1315.

15. John L. Skosey, "Some Basic Aspects of Bone Metabolism in Relation to Osteoporosis," *Medical Clinics of North America*, 54 (1979): 141–52.

16. H. P. Newton-John and D. Brian Morgan, "The Loss of Bone with Age, Osteoporosis, and Fractures," *Clinical Orthopaedics and Related Research*, 71 (1970): 229–52.

17. Jenifer Jowsey, "Osteoporosis: Dealing with a Crippling Bone Disease of the Elderly," *Geriatrics*, 32 (1977): 41–50; M. R. A. Khairi and C. Conrad Johnston, "What We Know–and Don't Know–about Bone Loss in the Elderly," *Clinical Orthopaedics and Related Research*, 33 (1978): 67–76; and Louis V. Avioli, "What to Do with 'Postmenopausal Osteoporosis'?," *American Journal of Medicine*, 65 (1978): 881–84.

18. Gordan and Vaughan, *Clinical Management*, 2, 4, 93.

19. Silvia Meema, Manzer L. Bunker, and H. Erik Meema, "Preventive Effect of Estrogen on Postmenopausal Bone Loss," *Archives of Internal Medicine*, 135 (1975): 1436–40; M. Edward Davis, Lawrence H. Lanzl, and Ann B. Cook, "Detection, Prevention, and Retardation of Menopausal Osteoporosis," *Obstetrics and Gynecology*, 36 (1970): 187–98; and Gilbert S. Gordan and Cynthia Vaughan, "Postmenopausal Osteoporosis," *Primary Care*, 1 (1974): 565–83.

20. Mary Wheeler, "Osteoporosis," *Medical Clinics of North America*, 60 (1976): 1222.

21. Nathan Kase, "Estrogens and the Menopause," *JAMA*, 227 (1974): 318–19; "Estrogen Therapy: The Dangerous Road to Shangri-La," *Consumer Reports*, Nov. 1976, 642–45.

22. Robert P. Heaney, "Estrogens and Postmenopausal Osteoporosis," *Clinical Obstetrics and Gynecology*, 19 (1976): 791–803. See also Heaney, "Estrogens and Osteoporosis [editorial]," *Western Journal of Medicine*, 125 (1976): 149–50.

23. Wheeler, "Osteoporosis," 1222–23; Sherry F. Queener and Norman H. Bell, "Calcitonin: A General Survey," *Metabolism*, 24 (1975): 555–67; J. Dequeker, "Bones and Ageing," *Annals of Rheumatic Diseases*, 34 (1975): 100–115; Louis V. Avioli, "Senile and Postmenopausal Osteoporosis," *Advances in Internal Medicine*, 21 (1976): 404–7; Jerome Targovnik, "Senile Osteoporosis," *Radiologic Clinics of North America*, 15 (1977): 289–92; John F. Aloia, I. Zanzi, A. Vaswani, K. Ellis, and S. H. Cohn, "Combination Therapy for Osteoporosis," *Metabolism*, 26 (1977): 787–92; and Stephen J. Marx, "Restraint in Use of High-Dose Fluorides to Treat Skeletal Disorders," *JAMA*, 240 (1978): 1630–31.

24. Uriel S. Barzel, "The Challenge of Osteoporosis," *Archives of Physical Medicine and Rehabilitation*, 52 (1971): 136.

25. Wheeler, "Osteoporosis," 1223.

26. Steven R. Cummings, J. L. Kelsey, M. C. Nevitt, and K. J. O'Dowd, "Epidemiology of Osteoporosis and Osteoporotic Fractures," *Epidemiologic Reviews*, 7 (1985): 198.

27. Herbert Fleisch, R. G. Russell, B. Simpson, and R. C. Mühlbauer, "Prevention by a Diphosphonate of Immobilization 'Osteoporosis' in Rats," *Nature*, 223 (1969): 211–12; Fleisch, Russell, and Marion D. Francis, "Diphosphonates Inhibit Hydroxypatite Dissolution In Vitro and Bone Resorption in Tissue Culture and In Vivo," *Science*, 165 (1969): 1262–64; Miguel E. Cabanela and Jenifer Jowsey, "The Effects of Phosphonates on Experimental Osteoporosis," *Calcified Tissue Research*, 8 (1971): 114–20; and Joseph M. Lane and Marvin E. Steinberg, "Inhibition of Disuse Osteoporosis by Diphosphonates," *Surgical Forum*, 23 (1972): 473–74.

28. The following discussion of the history of estrogen follows the authoritative study by Watkins, *Estrogen Elixir*.

29. Ibid., 32–36; and William H. Masters and Willard M. Allen, "Female Sex Hormone Replacement in the Aged Woman," *Journal of Gerontology*, 3 (1948): 183–90.

30. Watkins, *Estrogen Elixir*, 37–39.

31. Ibid., 39–44.

32. Ibid., 44–46.

33. Robert A. Wilson and Thelma N. Wilson, "The Fate of the Nontreated Postmenopausal Woman: A Plea for the Maintenance of Adequate Estrogen from Puberty to the Grave," *Journal of the American Geriatrics Society*, 11 (1963): 347–62.

34. Robert A. Wilson, *Feminine Forever* (New York: M. Evans, 1966), 15–27, 196, 201–3.

35. Watkins, *Estrogen Elixir*, 47–49, 69–70.

36. David R. Reuben, *Everything You Always Wanted to Know about Sex, but Were Afraid to Ask* (New York: David McKay, 1969), 288–300.

37. Watkins, *Estrogen Elixir*, 69–92.

38. Donald C. Smith, R. Prentice, D. J. Thompson, and W. L. Herrmann, "Association of Exogenous Estrogen and Endometrial Carcinoma," *New England Journal of Medicine*, 293 (1975): 1164–67; and Harry K. Ziel and William D. Finkle,

"Increased Risk of Endometrial Carcinoma among Users of Conjugated Estrogens," *New England Journal of Medicine*, 1167–70; Watkins, *Estrogen Elixir*, 93–102.

39. Noel S. Weiss, "Risks and Benefits of Estrogen Use," *New England Journal of Medicine*, 293 (1975): 1200–1202.

40. Watkins, *Estrogen Elixir*, 132–47, 162–64.

41. Martin M. Quigley and Charles R. Hammond, "Estrogen Replacement Therapy," *New England Journal of Medicine*, 301 (1979): 646–48.

42. "Estrogen Use and Postmenopausal Women," NIH Consensus Statement, Sept. 13–14, 1979, http://consensus.nih.gov/historical.htm; and Barbara Gastel, Joan Cornoni-Huntley, and Jacob A. Brody, "Estrogen Use and Postmenopausal Women: A Basis for Informed Decisions," *Journal of Family Practice*, 11 (1980): 851–60.

43. Watkins, *Estrogen Elixir*, 148–49.

44. Lila E. Nachtigall, Richard H. Nachtigall, Robert D. Nachtigall, and Mark E. Beckman, "Estrogen Replacement Therapy I: A 10-Year Prospective Study in the Relationship to Osteoporosis," *Obstetrics and Gynecology*, 53 (1979): 277–81; and L. Nachtigall, R. H. Nachtigall, R. D. Nachtigall, and Beckman, "Estrogen Replacement Therapy II: A Prospective Study in the Relationship to Carcinoma and Cardiovascular and Metabolic Problems," *Obstetrics and Gynecology*, 54 (1979): 74–79.

45. Nachtigall et al., "Estrogen Replacement Therapy I," 280.

46. For all intents and purposes, HRT and ERT were used interchangeably by researchers and clinicians, even though ERT referred to estrogen.

47. Watkins, *Estrogen Elixir*, 156–57.

48. Tom A. Hutchinson, Stanley M. Polansky, and Alvin R. Feinstein, "Post-Menopausal Oestrogens Protect Against Fractures of Hip and Distal Radius: A Case-Control Study," *Lancet*, 314 (Oct. 6, 1979): 705–9.

49. R. Lindsay, D. M. Hart, D. Purdie, M. M. Ferguson, A. S. Clark, and A. Kraszewski, "Comparative Effects of Oestrogen and a Progestogen on Bone Loss in Postmenopausal Women," *Clinical Science and Molecular Medicine*, 54 (1978): 193–95; Lindsay, Hart, A. MacLean, A. C. Clark, Kraszewski, and J. Garwood, "Bone Response to Termination of Oestrogen Treatment," *Lancet*, 311 (Jun. 24, 1978): 1325–27; and Lindsay, Hart, C. Forrest, and C. Baird, "Prevention of Spinal Osteoporosis in Oophorectomised Women," *Lancet*, 316 (Nov. 29, 1980): 1151–53.

50. G. Donald Whedon, "Recent Advances in Management of Osteoporosis," *Advances in Experimental Medicine and Biology*, 128 (1980): 604.

51. Hershel Jick, Alexander M. Walker, and Kenneth J. Rothman, "The Epidemic of Endometrial Cancer: A Commentary," *American Journal of Public Health*, 70 (1980): 264–67.

52. Beverly A. Mosher and Elizabeth M. Whelen, "Postmenopausal Estrogen Therapy: A Review," *Obstetrical and Gynecological Survey*, 36 (1981): 467–75.

53. Alvan R. Feinstein and Ralph I. Horwitz, "A Critique of the Statistical Evidence Associating Estrogens with Endometrial Cancer," *Cancer Research*, 38 (1978): 4001–5.

54. Lawrence G. Raisz, "Osteoporosis," *Journal of the American Geriatrics Society*, 30 (1982): 127–38.

55. Council on Scientific Affairs of the American Medical Association, "Estrogen Replacement in the Menopause," *JAMA*, 249 (1983): 359–61.

56. For examples of this ambivalence, see Noel S. Weiss, C. L. Ure, J. H. Ballard, A. R. Williams, and J. R. Daling, "Decreased Risk of Fractures of the Hip and Lower Forearm with Postmenopausal Use of Estrogen," *New England Journal of Medicine*, 303 (1980): 1195–98; Jean L. Marx, "Osteoporosis: New Help for Thinning Bones," *Science*, 207 (1980): 628–30; Thomas C. Vaughn and Charles B. Hammond, "Estrogen Replacement Therapy," *Clinical Obstetrics and Gynecology*, 24 (1981): 253–83; Brian L. Strom, "Are Estrogens Effective in Preventing Fractures from Postmenopausal Osteoporosis?," *Drug Therapeutics*, 1982, 67–80; Kenneth J. Ryan, "Postmenopausal Estrogen Use," *Annual Review of Medicine*, 33 (1982): 171–81; Lombardo F. Palma, "Postmenopausal Osteoporosis and Estrogen Therapy: Who Should be Treated?," *Journal of Family Practice*, 14 (1982): 355–59; Howard L. Judd, D. L. Meldrum, L. J. Deftos, and B. E. Henderson, "Estrogen Replacement Therapy: Indications and Complications," *Annals of Internal Medicine*, 98 (1983): 195–205; Claus Christiansen and Paul Rodbro, "Does Postmenopausal Bone Loss Respond to Estrogen Replacement Therapy Independent of Bone Loss Rate?," *Calcified Tissue International*, 35 (1983): 720–22; and Pedro J. Beauchamp and Berel Held, "Estrogen Replacement Therapy: Universal Remedy for the Postmenopausal Woman?," *Postgraduate Medicine*, 75 (1984): 42–53.

57. Cummings et al., "Epidemiology of Osteoporosis," 178.

58. Gilbert S. Gordan, "If Preventable Why Not Prevented?," *Western Journal of Medicine*, 133 (1980): 331–33. For other examples of support for ERT, see John F. Aloia, "Estrogen and Exercise in Prevention and Treatment of Osteoporosis," *Geriatrics*, 37 (1982): 81–85; Richard Warnich, Katsuhiko Yano, and John Vogel, "Postmenopausal Bone Loss at Multiple Skeletal Sites: Relationship to Estrogen Use," *Journal of Chronic Diseases*, 36 (1983): 781–90; and Albert W. Diddle and Ira Q. Smith, "Postmenopausal Osteoporosis: The Role of Estrogens," *Southern Medical Journal*, 77 (1984): 868–74.

59. R. Don Gambrell Jr., F. M. Massey, T. A. Castaneda, A. J. Ugenas, and C. A. Ricci, "Reduced Incidence of Endometrial Cancer among Postmenopausal Women Treated with Progestogens," *Journal of the American Geriatrics Society*, 27 (1979): 389–94; Gambrell, Massey, Castaneda, and A. W. Boddie, "Estrogen Therapy and Breast Cancer in Postmenopausal Women," *Journal of the American Geriatrics Society*, 28 (1980): 251–57; R. B. Greenblatt, Gambrell, and L. D. Stoddard, "The Protective Role of Progesterone in the Prevention of Endometrial Cancer," *Pathology, Research, and Practice*, 174 (1982): 297–318; Gambrell, Robert C. Maier, and Barbara I. Sanders, "Decreased Incidence of Breast Cancer in Postmenopausal Estrogen-Progestogen Users," *Obstetrics and Gynecology*, 62 (1983): 435–43; Gambrell, Carol A. Bagnell, and Robert B. Greenblatt, "Role of Estrogens and Progesterone in the Etiology and Prevention of Endometrial Cancer: Review," *American*

Journal of Obstetrics and Gynecology, 146 (1983): 696–707; and Gambrell, "Estrogen-Progestogen Therapy during Menopause," *Postgraduate Medicine*, 80 (1986): 261–67.

60. M. I. Whitehead, P. T. Townsend, J. Pryse-Davies, T. A. Ryder, and R. J. King, "Effects of Estrogens and Progestins on the Biochemistry and Morphology of the Postmenopausal Endometrium," *New England Journal of Medicine*, 305 (1981): 1599–1605; and Watkins, *Estrogen Elixir*, 153.

61. Trudy L. Bush, Linda D. Cowan, Elizabeth Barrett-Connor, Michael H. Criqui, John M. Karon, Robert B. Wallace, H. Al Tyroler, and Basil M. Rifkind, "Estrogen Use and All-Cause Mortality: Preliminary Results from the Lipid Research Clinics Program Follow-Up Study," *JAMA*, 249 (1983): 903–6.

62. "Osteoporosis," NIH Consensus Development Conference Statement, Apr. 2–4, 1984, http://consensus.nih.gov/historical.htm; and "Osteoporosis," *JAMA*, 252 (1984): 799–802.

63. Watkins, *Estrogen Elixir*, 164–66.

64. Ibid., 167.

65. Ibid., 178–79.

66. Winnifred Berg Cutler, Celso-Ramón Garcia, and David A. Edwards, *Menopause: A Guide for Women and the Men Who Love Them* (New York: W. W. Norton, 1983), 106–21; and Cutler and Garcia, *The Medical Management of Menopause and Premenopause* (Philadelphia: J. B. Lippincott, 1984).

67. M. Notelovitz and P. van Keep, eds., *The Climacteric in Perspective: Proceedings of the Fourth International Congress on the Menopause, Held at Lake Buena Vista, Florida, October 28–November 2, 1984* (Hingham, MA: MTP Press, 1986), 19–21.

68. Morris Notelovitz and Marsha Ware, *Stand Tall!: The Informed Woman's Guide to Preventing Osteoporosis* (Gainesville, FL: Triad, 1982), 15, 112–25.

69. Louis V. Avioli, ed., *The Osteoporotic Syndrome: Detection, Prevention, and Treatment*, 1st ed. (New York: Grune & Stratton, 1983), viii, 71, 129, 140.

70. Louis V. Avioli, ed., *The Osteoporotic Syndrome: Detection, Prevention, and Treatment*, 2nd ed. (New York: Grune & Stratton, 1987), 91–107.

71. David F. Fardon, *Osteoporosis: Your Head Start on the Prevention and Treatment of Brittle Bones* (New York: Macmillan, 1985), 8–10, 80–99, 248, and elsewhere; and pers. comm., David F. Fardon to Gerald N. Grob, Jul. 5, 2012.

72. Lila Nachtigall, with Joan Heilman, *The Lila Nachtigall Report* (New York: Putnam, 1977).

73. Lila Nachtigall and Joan Rattner Heilman, *Estrogen: The Facts Can Change Your Life* (New York: Harper & Row, 1986), 199–200 and elsewhere.

74. Watkins, *Estrogen Elixir*, 184.

75. Lila Nachtigall, Robert D. Nachtigall, and Joan Rattner Heilman, *What Every Woman Should Know: Staying Healthy after 40* (New York: Warner Books, 1995).

Chapter 4. Popularizing a Diagnosis

1. *Wall Street Journal*, Jun. 15, 1995.

2. U.S. Senate, 99th Congress, 1st Session, *Osteoporosis: Hearing Before the Sub-*

committee on Aging of the Committee on Labor and Human Resources . . . Reviewing the Diagnosis and Treatment of Osteoporosis, June 20, 1985 (Washington, DC: U.S. Government Printing Office, 1985), 137–43; and "Founding and Milestones," National Osteoporosis Foundation, www.nof.org/aboutnof/foundingandmilestones/.

3. U.S. Senate, *Hearing Before the Subcommittee on Aging*, 1–9.

4. Ibid., 10–13.

5. Ibid., 16–38.

6. Takuo Fujita, "Calcium and Aging [editorial]," *Calcified Tissue International*, 37 (1985): 1–2. See also James M. Burnell, D. J. Baylink, C. H. Chestnut III, and E. J. Teubner, "The Role of Skeletal Calcium Deficiency in Postmenopausal Osteoporosis," *Calcified Tissue International*, 38 (1986): 187–92; and Herta Spencer, L. Kramer, M. Lesniak, M. De Bartolo, C. Norris, and D. Osis, "Calcium Requirements in Humans: Report of Original Data and a Review," *Clinical Orthopaedics and Related Research*, 184 (1984): 270–80.

7. U.S. Senate, *Hearing Before the Subcommittee on Aging*, 39–86.

8. Ibid., 87–113, 125–36.

9. Ibid., 144–64, 170–79.

10. Barron H. Lerner, *The Breast Cancer Wars: Hope, Fear, and the Pursuit of a Cure in Twentieth-Century America* (New York: Oxford University Press, 2001), 141, 223, and elsewhere.

11. U.S. Senate, *Hearing Before the Subcommittee on Aging*, 166–69.

12. Lawrence G. Raisz and JoAnne Smith, "Prevention and Therapy of Osteoporosis," *Rational Drug Therapy*, 19 (1985): 1–6.

13. For a discussion of risk factors, see William G. Rothstein, *Public Health and the Risk Factor: A History of an Uneven Medical Revolution* (Rochester, NY: University of Rochester Press, 1993); and Gerald N. Grob, *The Deadly Truth: A History of Disease in America* (Cambridge, MA: Harvard University Press, 2002), 248–59.

14. Lawrence G. Raisz and A. Johannesson, "Pathogenesis, Prevention, and Therapy of Osteoporosis," *Journal of Medicine*, 15 (1984): 272–73.

15. Steven R. Cummings, Jennifer L. Kelsey, Michael C. Nevitt, and Kenneth J. O'Dowd, "Epidemiology of Osteoporosis and Osteoporotic Fractures," *Epidemiologic Reviews*, 7 (1985): 178–208.

16. Steven R. Cummings, "Are Patients with Hip Fractures More Osteoporotic?: Review of the Evidence," *American Journal of Medicine*, 78 (1985): 487–94.

17. L. Joseph Melton III and B. Lawrence Riggs, "Risk Factors for Injury after a Fall," *Clinics in Geriatric Medicine*, 1 (1985): 525–39; Melton, "Epidemiology of Fractures," in *Osteoporosis: Etiology, Diagnosis, and Management*, 1st ed., ed. Riggs and Melton (New York: Raven Press, 1988), 135; and Melton, Edmund Y. S. Chao, and Joseph Lane, "Biomechanical Aspects of Fractures," in *Osteoporosis: Etiology, Diagnosis, and Management*, 1st ed., ed. Riggs and Melton, 125–26.

18. John F. Aloia, S. H. Cohn, A. Vaswani, J. K. Yeh, K. Yuen, and K. Ellis, "Risk Factors for Postmenopausal Osteoporosis," *American Journal of Medicine*, 78 (1985): 95–100.

19. R. Wootton, E. Bryson, U. Elsasser, H. Freeman, J. R. Green, R. Hesp, E. A. Hudson, L. Klenerman, T. Smith, and J. Zanelli, "Risk Factors for Fractured Neck of Femur in the Elderly," *Age and Ageing*, 11 (1982): 160–68.

20. A. Michael Parfitt, "Dietary Risk Factors for Age-Related Bone Loss and Fractures," *Lancet*, 322 (Nov. 19, 1983): 1181–85.

21. "Risk Factors in Postmenopausal Osteoporosis," *Lancet*, 325 (Jun. 15, 1985): 1370–72.

22. Jacob A. Brody, M. E. Farmer, and L. R. White, "Absence of Menopausal Effect on Hip Fracture Occurrence in White Females," *American Journal of Public Health*, 74 (1984): 1397–98; and B. Lawrence Riggs, H. W. Wahner, E. Seeman, K. P. Offord, W. L. Dunn, R. B. Mazess, K. A. Johnson, and L. J. Melton III, "Changes in Bone Mineral Density of the Proximal Femur and Spine with Aging: Differences between the Postmenopausal and Senile Osteoporosis Syndromes," *Journal of Clinical Investigation*, 70 (1982): 716–23.

23. Mary E. Farmer, L. R. White, J. A. Brody, and K. R. Bailey, "Race and Sex Differences in Hip Fracture Incidence," *American Journal of Public Health*, 74 (1984): 1374–80.

24. David F. Fardon, *Osteoporosis: Your Head Start on the Prevention and Treatment of Brittle Bones* (New York: Macmillan, 1985), 19.

25. American Child Health Association, *Physical Defects: The Pathway to Correction* (New York: American Child Health Association, 1934), 80–96.

26. For a discussion, see Stanley J. Reiser, "The Emergence of the Concept of Screening for Disease," *Milbank Memorial Fund Quarterly*, 56 (1978): 403–25.

27. B. E. C. Nordin, "The Definition and Diagnosis of Osteoporosis," *Calcified Tissue International*, 40 (1987): 57–58.

28. Steven R. Cummings, "Should Perimenopausal Women Be Screened for Osteoporosis?," *Annals of Internal Medicine*, 104 (1986): 817.

29. Vincent R. Sites, "Involutional Osteoporosis [letter]," *New England Journal of Medicine*, 316 (1987): 215.

30. Cummings, "Should Perimenopausal Women Be Screened?," 817.

31. Ibid., 817–23.

32. Michele F. Bellantoni and Marc R. Blackman, "Osteoporosis: Diagnostic Screening and Its Place in Current Care," *Geriatrics*, 43 (1988): 63–68.

33. Susan Ott, "Should Women Get Screening Bone Mass Measurements? [editorial]," *Annals of Internal Medicine*, 104 (1986): 874–76.

34. B. Lawrence Riggs and L. Joseph Melton III, "Involutional Osteoporosis," *New England Journal of Medicine*, 314 (1985): 1676–86.

35. "Bone Mineral Densitometry," *Annals of Internal Medicine*, 109 (1988): 846.

36. Ferris M. Hall, Michael A. Davis, and Daniel T. Baran, "Bone Mineral Screening for Osteoporosis," *New England Journal of Medicine*, 316 (1987): 212–14. For letters responding to this article, see *New England Journal of Medicine*, 317 (1987): 315–16.

37. "Bone Mineral Densitometry: Society Responds to HCFA and Blue Cross/Blue Shield Subcommittee," *Journal of Nuclear Medicine*, 26 (1985): 115–17.

38. David L. Sartoris and Donald Resnick, "Densitometric Screening for Osteoporosis: A Radiologist's Perspective," *American Journal of Roentgenology*, 147 (1986): 1329–30.

39. National Osteoporosis Foundation, "Clinical Indications for Bone Mass Measurements," *Journal of Bone and Mineral Research*, 4, Suppl. 2 (1989), 1.

40. The full report can be found in ibid., 1–28.

41. Bruce E. Johnson, B. Lucasey, R. G. Robinson, and B. P. Lukert, "Contributing Diagnoses in Osteoporosis: The Value of a Complete Medical Evaluation," *Archives of Internal Medicine*, 149 (1989): 1069–72.

42. B. Lawrence Riggs, H. W. Wahner, L. J. Melton III, L. S. Richelson, H. L. Judd, and K. P. Offord, "Rates of Bone Loss in the Appendicular and Axial Skeletons of Women: Evidence of Substantial Bone Loss before Menopause," *Journal of Clinical Investigation*, 77 (1986): 1487–91.

43. Ignac Fogelman, "An Evaluation of the Contribution of Bone Mass Measurements to Clinical Practice," *Seminars in Nuclear Medicine*, 19 (1989): 62–68.

44. Robert Lindsay, "Osteoporosis: An Updated Approach to Prevention and Management," *Geriatrics*, 44 (1989): 45.

45. "Consensus Development Conference: Prophylaxis and Treatment of Osteoporosis," *British Medical Journal*, 295 (1987): 954–55.

46. Tony Smith, [letter], *British Medical Journal*, 295 (1987): 872; and B. E. C. Nordin, A. G. Need, H. A. Morris, and M. Horowitz, [letter], *British Medical Journal*, 295 (1987): 1276–77.

47. Elizabeth S. Watkins, *The Estrogen Elixir: A History of Hormone Replacement Therapy in America* (Baltimore: Johns Hopkins University Press, 2007), 187–93.

48. Ibid., 202–3.

49. Ibid., 63–67, 102–3, 247–48; and Barbara Seaman, *The Greatest Experiment Ever Performed on Women: Exploding the Estrogen Myth* (New York: Hyperion, 2003), 47–60.

50. Richard J. Haber, "Should Postmenopausal Women Be Given Estrogen?," *Western Journal of Medicine*, 142 (1985): 672–77.

51. Leonore C. Huppert, "Hormonal Replacement Therapy: Benefits, Risks, Doses," *Medical Clinics of North America*, 71 (1987): 26; Robert Lindsay, "Estrogens in Prevention and Treatment of Osteoporosis," in *The Osteoporotic Syndrome: Detection, Prevention, and Treatment*, 2nd ed., ed. Louis V. Avioli (Orlando, FL: Grune & Stratton, 1987), 91–107; and Lindsay, "Estrogen Therapy in the Prevention and Management of Osteoporosis," *American Journal of Obstetrics and Gynecology*, 156 (1987): 1351.

52. Edward G. Lufkin, P. C. Carpenter, S. J. Ory, G. D. Malkasian, and J. H. Edmonson, "Estrogen Replacement Therapy: Current Recommendations," *Mayo Clinic Proceedings*, 63 (1988): 453–60.

53. Huppert, "Hormonal Replacement Therapy," 36.

54. Benjamin F. Byrd Jr., John C. Burch, and William K. Vaughn, "The Impact of Long Term Estrogen Support after Hysterectomy," *Annals of Surgery*, 185 (1977): 574–79.

55. Trudy L. Bush, Linda D. Cowan, Elizabeth Barrett-Connor, Michael H. Criqui, John M. Karon, Robert B. Wallace, H. Al Tyroler, and Basil M. Rifkind, "Estrogen Use and All-Cause Mortality: Preliminary Results from the Lipid Research Clinics Program Follow-Up Study," *JAMA*, 249 (1983): 903–6.

56. William A. Peck and Louis V. Avioli, *Osteoporosis: The Silent Thief* (Glenview, IL: Scott, Foresman, 1988), 96–108.

57. B. Lawrence Riggs, "Practical Management of the Patient with Osteoporosis," in *Osteoporosis: Etiology, Diagnosis, and Management*, 1st ed., ed. Riggs and Melton, 481–90.

58. Louis V. Avioli, "The Calcium Controversy and the Recommended Dietary Allowance," in *Osteoporotic Syndrome*, 2nd ed., ed. Avioli, 57–66.

59. J. Chris Gallagher, "Drug Therapy of Osteoporosis: Calcium, Estrogen, and Vitamin D," in *Osteoporosis: Etiology, Diagnosis, and Management*, 1st ed., ed. Riggs and Melton, 389–93.

60. Richard W. Barth and Joseph M. Lane, "Osteoporosis," *Orthopedic Clinics of North America*, 19 (1988): 845.

61. Raisz and Smith, "Prevention and Therapy of Osteoporosis," 5; Carlo Gennari and Louis V. Avioli, "Calcitonin Therapy in Osteoporosis," in *Osteoporotic Syndrome*, 1st ed., ed. Avioli (New York: Grune & Stratton, 1983), 121–42; and Charles H. Chestnut III, "Drug Therapy, Calcitonin, Bisphosphonates, Anabolic Steroids, and HPTH (1–34)," in *Osteoporosis: Etiology, Diagnosis, and Management*, 1st ed., ed. Riggs and Melton, 404–7.

62. Clayton Rich, John Ensinck, and Peter Ivanovich, "The Effects of Sodium Fluoride on Calcium Metabolism of Subjects with Metabolic Bone Diseases," *Journal of Clinical Investigation*, 41 (1964): 545–56.

63. David J. Baylink and Daniel S. Bernstein, "The Effects of Fluoride Therapy on Metabolic Bone Disease," *Clinical Orthopaedics and Related Research*, 55 (1967): 51–85.

64. Louis V. Avioli, "The Management of the Geriatric Osteoporotic Woman," in *Osteoporotic Syndrome*, 2nd ed., ed. Avioli, 151; Erik F. Erikson, Stephen F. Hodgson, and B. Lawrence Riggs, "Treatment of Osteoporosis with Sodium Fluoride," in *Osteoporosis: Etiology, Diagnosis, and Management*, 1st ed., ed. Riggs and Melton, 413–32; and Louis V. Avioli, "Therapy with Bisphosphonates, Sodium Fluoride, Thiazides, and Vitamin D Metabolites," in *Osteoporotic Syndrome*, 3rd ed., ed. Avioli (New York: Wiley-Liss, 1993), 160–62.

65. B. Lawrence Riggs, S. F. Hodgson, W. M. O'Fallon, E. Y. Chao, H. W. Wahner, J. M. Muhs, S. L. Cedel, and L. J. Melton III, "Effect of Fluoride Treatment on the Fracture Rate in Postmenopausal Women with Osteopo-

rosis," *New England Journal of Medicine*, 322 (1990): 802–9; and M. Kleerekoper, E. L. Peterson, D. A. Nelson, E. Phillips, M. A. Schork, B. C. Tilley, and A. M. Parfitt, "A Randomized Trial of Sodium Fluoride as a Treatment for Postmenopausal Osteoporosis," *Osteoporosis International*, 1 (1991): 155–61.

66. Robert P. Heaney and Paul D. Saville, "Etidronate Disodium in Postmenopausal Osteoporosis," *Clinical Pharmacology and Therapeutics*, 20 (1976): 593–604.

67. Chestnut, "Drug Therapy, Calcitonin, Bisphosphonates," in *Osteoporosis: Etiology, Diagnosis, and Management*, 1st ed., ed. Riggs and Melton, 407–9.

68. Peck and Avioli, *Osteoporosis: The Silent Thief*, viii, iii–ix, 7.

69. Avioli, ed., *Osteoporotic Syndrome*, 2nd ed. (1983); 3rd. ed. (1993); and 4th ed. (San Diego: Academic Press, 2000).

70. Riggs and Melton, *Osteoporosis: Etiology, Diagnosis, and Management*, 1st ed.; and 2nd ed. (Philadelphia: Lippincott-Raven, 1993).

71. Roberto Pacifici and Louis V. Avioli, "Effects of Aging on Bone Structure and Metabolism," in *Osteoporotic Syndrome*, 3rd ed., ed. Avioli, 1–2.

72. Charles E. Dent, "Keynote Address: Problems in Metabolic Bone Disease," in *Clinical Aspects of Metabolic Bone Disease: Proceedings of the International Symposium on Clinical Aspects of Metabolic Bone Disease; Henry Ford Hospital, Detroit, Michigan, June 16–19, 1972*, ed. Boy Frame, A. M. Parfitt, and Howard Duncan (Amsterdam: Excerpta, 1973), 5–6.

Chapter 5. Internationalizing Osteoporosis

1. Robert P. Heaney, "Osteoporosis at the End of the Century," *Western Journal of Medicine*, 154 (1991): 106–7.

2. L. Joseph Melton III, "Hip Fractures: A Worldwide Problem Today and Tomorrow," *Bone*, 14, Suppl. 1 (1993): S1–S8.

3. Robert Lindsay and Pierre J. Meunier, "Editorial," *Osteoporosis International*, 1 (1990): 1.

4. "Consensus Development Conference: Prophylaxis and Treatment of Osteoporosis," *British Medical Journal*, 295 (1987): 914–15.

5. "Proceedings of a Symposium: Consensus Development Conference on Osteoporosis," *American Journal of Medicine*, 91, Suppl. 5B (1991): 1S–68S. Another summary article claimed that 4,000 researchers and clinicians were at the conference. See "Meeting Highlights: Third International Symposium on Osteoporosis," *Orthopedic Review*, 20 (1991): 284–91.

6. Charles E. Dent, "Keynote Address: Problems in Metabolic Bone Disease," in *Clinical Aspects of Metabolic Bone Disease: Proceedings of the International Symposium on Clinical Aspects of Metabolic Bone Disease; Henry Ford Hospital, Detroit, Michigan, June 16–19, 1972*, ed. Boy Frame, A. M. Parfitt, and Howard Duncan (Amsterdam: Excerpta, 1973), 5–6.

7. Charles H. Chestnut III, "Theoretical Overview: Bone Development, Peak Bone Mass, Bone Loss, and Fracture Risk," *American Journal of Medicine*, 91, Suppl. 5B (1991): 2S–4S.

8. John A. Eisman, P. N. Sambrook, P. J. Kelly, and N. A. Pocock, "Exercise and Its Interaction with Genetic Influences in the Determination of Bone Mineral Density," *American Journal of Medicine*, 91, Suppl. 5B (1991): 5S–9S.

9. Robert Lindsay, "Estrogens, Bone Mass, and Osteoporotic Fracture," *American Journal of Medicine*, 91, Suppl. 5B (1991): 10S–13S.

10. Steven R. Cummings, "Evaluating the Benefits and Risks of Postmenopausal Hormone Therapy," *American Journal of Medicine*, 91, Suppl. 5B (1991): 14S–18S.

11. Jean-Yves Reginster, "Effect of Calcitonin on Bone Mass and Fracture," *American Journal of Medicine*, 91, Suppl. 5B (1991): 19S–22S.

12. Robert P. Heaney, "Effect of Calcium on Skeletal Development, Bone Loss, and Risk of Fractures," *American Journal of Medicine*, 91, Suppl. 5B (1991): 23S–28S.

13. The following were also published in the *American Journal of Medicine*, 91, Suppl. 5B (1991): John A. Kanis, "The Restoration of Skeletal Mass: A Theoretic Overview," 29S–36S; B. Lawrence Riggs, "Treatment of Osteoporosis with Sodium Fluoride or Parathyroid Hormone," 37S–41S; A. Michael Parfitt, "Use of Bisphosphonates in the Prevention of Bone Loss and Fractures," 42S–46S; C. Conrad Johnson and Charles W. Slemenda, "Risk Prediction in Osteoporosis: A Theoretic Overview," 47S–48S; Harry K. Genant, Kenneth G. Faulkner, and Claus-Christian Gloer, "Measurement of Bone Mineral Density: Current Status," 49S–53S; Richard D. Wasnich, "Bone Mass Measurements in Diagnosis and Assessment of Therapy," 54S–58S; P. D. Delmas, "Biochemical Markers of Bone Turnover: Methodology and Clinical Use in Osteoporosis," 59S–63S; and Bente Juel Riis, "Biochemical Markers of Bone Turnover in Diagnosis and Assessment of Therapy," 64S–68S. Also see Pierre J. Meunier, "Restoration of Skeletal Mass in Secondary Osteoporosis," *Osteoporosis International*, 1 (1991): 123.

14. "Consensus Development Conference: Prophylaxis and Treatment of Osteoporosis," *American Journal of Medicine*, 90 (1991): 107–10.

15. Roger Smith, "Osteoporosis at the Tivoli," *Journal of Bone and Joint Surgery* (British ed.), 73 (1991): 525–26.

16. Sherry Sherman and Evan C. Hadley, "Aging and Bone Quality: An Underexplored Frontier," *Calcified Tissue International*, 53, Suppl. 1 (1993): S1.

17. A. Michael Parfitt, "Overview of Fracture Pathogenesis," *Calcified Tissue International*, 53, Suppl. 1 (1993): S2.

18. Robert P. Heaney, "Is There a Role for Bone Quality in Fragility Fractures?," *Calcified Tissue International*, 53, Suppl. 1 (1993): S3–S6.

19. Susan M. Ott, "When Bone Mass Fails to Predict Bone Failure," *Calcified Tissue International*, 53, Suppl. 1 (1993): S7–S13.

20. Philip D. Ross, James W. Davis, and Richard D. Wasnich, "Bone Mass and Beyond: Risk Factors for Fractures," *Calcified Tissue International*, 53, Suppl. 1 (1993): S134–S138.

21. Cyrus Cooper, "The Epidemiology of Fragility Fractures: Is There a Role for Bone Quality?," *Calcified Tissue International*, 53, Suppl. 1 (1993): S23–S26; and Christine M. Schnitzler, "Bone Quality: A Determinant for Certain Risk Factors for Bone Fragility," *Calcified Tissue International*, 53, Suppl. 1 (1993): S27–S31.

22. "Conclusion: Research Prospects and Initiatives," *Calcified Tissue International*, 53, Suppl. 1 (1993): S176–S180.

23. Reinhard Ziegler, "Osteoporosis Will Never Be a Disease of the Past," *Osteoporosis International*, 4, Suppl. 1 (1994): S3–S4.

24. Gideon A. Rodan, "Good Hope for Making Osteoporosis a Disease of the Past," *Osteoporosis International*, 4, Suppl. 1 (1994): S5–S6.

25. The following were also published in *Osteoporosis International*, 4, Suppl. 1 (1994): J.-Ph. Bonjour, G. Theintz, F. Law, D. Slosman, and R. Rizzoli, "Peak Bone Mass," S7–S13; E. Seeman, "Reduced Bone Density in Women with Fractures: Contribution of Low Peak Bone Density and Rapid Bone Loss," S15–S25; S. Adami, "Optimizing Peak Bone Mass: What Are the Therapeutic Possibilities?," S27–S30; D. T. Baran, "Magnitude and Determinants of Premenopausal Bone Loss," S31–S34; B. J. Riis, "Premenopausal Bone Loss: Fact or Artifact?," S35–S37; R. Lindsay, "Bone Mass Measurement for Premenopausal Women," S39–S41; C. C. Johnston and C. W. Slemenda, "Peak Bone Mass, Bone Loss, and Risk of Fracture," S43–S45; C. Christiansen, "Postmenopausal Bone Loss and the Risk of Osteoporosis," S47–S51; L. G. Raisz, "Assessment of the Risk of Osteoporosis at the Menopause: Therapeutic Consequences," S53–S57; and J. A. Kanis and S. Adami, "Bone Loss in the Elderly," S59–S65.

26. P. J. Meunier, M. C. Chapuy, M. E. Arlot, P. D. Delmas, and F. Duboeuf, "Can We Stop Bone Loss and Prevent Hip Fractures in the Elderly?," *Osteoporosis International*, 4, Suppl. 1 (1994): S71–S76.

27. S. R. Cummings and M. C. Nevitt, "Non-Skeletal Determinants of Fractures: The Potential Importance of the Mechanics of Falls," *Osteoporosis International*, 4, Suppl. 1 (1994): S67–S70.

28. World Health Organization, *Assessment of Fracture Risk and Its Application to Screening for Postmenopausal Osteoporosis: Report of a WHO Study Group*, WHO Technical Report No. 843 (Geneva: World Health Organization, 1994), 1–2.

29. Ibid., 5–6.

30. See esp. Jeremy A. Greene, *Prescribing by Numbers: Drugs and the Definition of Disease* (Baltimore: Johns Hopkins University Press, 2007).

31. World Health Organization, *Assessment of Fracture Risk*, 94–101.

32. John A. Kanis, "Assessment of Fracture Risk and Its Application to Screening for Postmenopausal Osteoporosis: Synopsis of a WHO Report," *Osteoporosis International*, 4 (1994): 368–81.

33. John A. Kanis, Jean-Pierre Devogelaer, and Carlo Gennari, "Practical Guide for the Use of Bone Mineral Measurements in the Assessment of Treatment of Osteoporosis: A Position Paper of the European Foundation for Osteoporosis and Bone Disease," *Osteoporosis International*, 6 (1996): 256–61.

34. World Health Organization, *Guidelines for Preclinical Evaluation and Clinical Trials in Osteoporosis* (Geneva: World Health Organization, 1998), 1.

35. Ibid., 9–12; J.-P. Bonjour, P. Ammann, and R. Rizzoli, "Importance of Preclinical Studies in the Development of Drugs for the Treatment of Osteoporosis: A

Review Related to the 1998 WHO Guidelines," *Osteoporosis International*, 9 (1999): 380–81.

36. Bonjour et al., "Importance of Preclinical Studies," 379–93.

37. World Health Organization, *Guidelines for Preclinical Evaluation*, 24; Bonjour et al., "Importance of Preclinical Studies," 379–93.

38. World Health Organization, *Guidelines for Preclinical Evaluation*, 24; Bonjour et al., "Importance of Preclinical Studies," 383.

39. Harry K. Genant, C. Cooper, G. Poor, I. Reid, G. Ehrlich, J. Kanis, B. E. Nordin, et al., "Interim Report and Recommendations of the World Health Organization Task Force for Osteoporosis," *Osteoporosis International*, 10 (1999): 259–64.

40. D. M. Black, L. Palermo, H. Genant, and S. Cummings, "Four Reasons to Avoid the Use of BMD T-Scores in Treatment Decisions for Osteoporosis," *Journal of Bone and Mineral Research*, 11, Suppl. S1 (1996): S118.

41. R. D. Wasnich, "Consensus and the T-Score Fallacy," *Clinical Rheumatology*, 16 (1997): 337–39.

42. T. Sandor, D. Felsenberg, and E. Brown, "Comments on the Hypotheses Underlying Fracture Risk Assessment in Osteoporosis as Proposed by the World Health Organization," *Calcified Tissue International*, 64 (1999): 267–70.

43. J. A. Kanis, O. Johnell, A. Oden, B. Jonsson, C. De Laet, and A. Dawson, "Risk of Hip Fracture According to the World Health Organization Criteria for Osteopenia and Osteoporosis," *Bone*, 27 (2000): 585–90.

44. Paul Lips, "Epidemiology and Predictors of Fractures Associated with Osteoporosis," *American Journal of Medicine*, 103, 2A (1997): 3S–11S.

45. E. Allander, B. I. B. Lindahl, and the MEDOS Study Group, "The Mediterranean Osteoporosis Study (MEDOS): Theoretical and Practical Issues of a Major International Project on Hip Fracture Epidemiology," *Bone*, 14, Suppl. 1 (1993): S37–S43; and G. P. Lyritis and the MEDOS Study Group, "Epidemiology of Hip Fracture: The MEDOS Study," *Osteoporosis International*, 6, Suppl. 3 (1996): S11–S15.

46. Olof Johnell, Bo Gullberg, John A. Kanis, Erik Allander, Lars Elffors, Jan Dequeker, Guzin Dilsen, et al., "Risk Factors for Hip Fracture in European Women: The MEDOS Study," *Journal of Bone and Mineral Research*, 10 (1995): 1802–15.

47. Eliot Marshall, "Big Science Enters the Clinic," *Science*, 260 (1993): 744–47; and Elizabeth S. Watkins, *The Estrogen Elixir: A History of Hormone Replacement in America* (Baltimore: Johns Hopkins University Press, 2007), 226–27.

48. C. Cooper, T. O'Neill, and A. Silman, "The Epidemiology of Vertebral Fractures: European Vertebral Osteoporosis Study Group," *Bone*, 14, Suppl. 1 (1993): S89–S97; T. W. O'Neill, D. Felsenberg, J. Varlow, Cooper, J. A. Kanis, and A. J. Silman, "The Prevalence of Vertebral Deformity in European Men and Women: The European Vertebral Osteoporosis Study," *Journal of Bone and Mineral Research*, 11 (1996): 1010–18; and J. Dequeker, J. Pearson, J. Reeve, M. Henley, J. Bright, D. Felsenberg, W. Kalender, et al., "Dual X-Ray Absorptiometry—Cross-Calibration and Normative Reference Ranges for the Spine: Results of a European Community Concerted Action," *Bone*, 17 (1995): 247–54.

49. J. Reeve, on behalf of the EPOS Study Group, "The European Prospective Osteoporosis Study," *Osteoporosis International*, 6, Suppl. 3 (1996): S16–S19.

50. EPOS Study Group, "Incidence of Vertebral Fracture in Europe: Results from the European Prospective Osteoporosis Study (EPOS)," *Journal of Bone and Mineral Research*, 17 (2002): 716–24.

51. D. K. Roy, T. W. O'Neill, J. D. Finn, M. Lunt, A. J. Silman, D. Felsenberg, G. Armbrecht, et al., "Determinants of Incident Vertebral Fracture in Men and Women: Results from the European Prospective Osteoporosis Study (EPOS)," *Osteoporosis International*, 14 (2003): 19–26.

52. S. Kaptoge, L. I. Benevolenskaya, A. K. Bhalla, J. B. Cannata, S. Boonen, A. J. Falch, D. Felsenberg, et al., "Low BMD Is Less Predictive than Reported Falls for Future Limb Fractures in Women across Europe: Results from the European Prospective Osteoporosis Study," *Bone*, 36 (2005): 387–98.

53. A. A. Ismail, S. R. Pye, W. C. Cockerill, M. Lunt, A. J. Silman, J. Reeve, D. Banzer, et al., "Incidence of Limb Fractures across Europe: Results from the European Prospective Osteoporosis Study (EPOS)," *Osteoporosis International*, 13 (2002): 565–71.

54. E. Siris, P. Miller, E. Barrett-Connor, T. Abbott, L. Sherwood, and M. Berger, "Design of NORA, the National Osteoporosis Risk Assessment Program: A Longitudinal US Registry of Postmenopausal Women," *Osteoporosis International*, 8, Suppl. 1 (1998): S62–S69.

55. Ibid.; Charles H. Chestnut III, "Osteoporosis: An Underdiagnosed Disease [editorial]," *JAMA*, 286 (2001): 2865–66.

56. Ethel S. Siris, P. D. Miller, E. Barrett-Connor, K. G. Faulkner, L. E. Wehren, T. A. Abbott, M. L. Berger, A. C. Santora, and L. M. Sherwood, "Identification and Fracture Outcomes of Undiagnosed Low Bone Mineral Density in Postmenopausal Women: Results from the National Osteoporosis Risk Assessment," *JAMA*, 286 (2001): 2815–22.

57. Chestnut, "Osteoporosis: An Underdiagnosed Disease," 2866.

58. Paul D. Miller, E. S. Siris, E. Barrett-Connor, K. G. Faulkner, L. E. Wehren, T. A. Abbott, Y. T. Chen, M. L. Berger, A. C. Santora, and L. M. Sherwood, "Prediction of Fracture Risk in Postmenopausal White Women with Peripheral Bone Densitometry: Evidence from the National Osteoporosis Risk Assessment," *Journal of Bone and Mineral Research*, 17 (2002): 2222–30.

59. Ethel S. Siris, S. K. Brenneman, P. D. Miller, E. Barratt-Connor, Y. T. Chen, L. M. Sherwood, and T. A. Abbott, "Predictive Value of Low BMD for 1-Year Fracture Outcomes Is Similar for Postmenopausal Women Ages 50–64 and 65 and Older: Results from the National Osteoporosis Risk Assessment (NORA)," *Journal of Bone and Mineral Research*, 19 (2004): 1215–20.

60. E. Barrett-Connor, T. W. Weiss, C. A. McHorney, P. D. Miller, and E. S. Siris, "Predictors of Falls among Postmenopausal Women: Results from the National Osteoporosis Risk Assessment (NORA)," *Osteoporosis International*, 20 (2009): 715–22.

61. Ibid.

62. J. H. Sheldon, "On the Natural History of Falls in Old Age," *British Medical Journal*, 2 (Dec. 10, 1960): 1685–90.

63. John M. Aitken, "Relevance of Osteoporosis in Women with Fracture of the Femoral Neck," *British Medical Journal*, 298 (1984): 597–601.

64. Steven R. Cummings, "Are Patients with Hip Fractures More Osteoporotic? Review of the Evidence," *American Journal of Medicine*, 78 (1985): 487–94.

65. L. Joseph Melton III, "Epidemiology of Fractures," in *Osteoporosis: Etiology, Diagnosis, and Management*, 1st ed., ed. B. Lawrence Riggs and Melton (New York: Raven Press, 1988), 136.

66. F. H. Hooven, J. D. Adachi, S. Adami, S. Boonen, J. Compston, C. Cooper, P. Delmas, et al., "The Global Longitudinal Study of Osteoporosis in Women (GLOW): Rationale and Study Design," *Osteoporosis International*, 20 (2009): 1107–16.

67. E. S. Siris, S. Gehlbach, J. D. Adachi, S. Boonen, R. D. Chapuriat, J. E. Compston, C. Cooper, et al., "Failure to Perceive Increased Risk of Fracture in Women 55 Years and Older: The Global Longitudinal Study of Osteoporosis in Women (GLOW)," *Osteoporosis International*, 22 (2011): 27–35.

68. Juliet E. Compston, N. B. Watts, R. Chapuriat, C. Cooper, S. Boonen, S. Greenspan, J. Pfeilschifter, et al., "Obesity Is Not Protective against Fracture in Postmenopausal Women: GLOW," *American Journal of Medicine*, 124 (2011): 1043–50.

69. Silvano Adami, G. Isaia, G. Luisetto, S. Minisola, L. Sinigaglia, R. Gentilella, D. Agnusdei, N. Iori, and R. Nuti, on behalf of the ICARO [Incidence and Characterization of Inadequate Clinical Responders in Osteoporosis] Study Group, "Fracture Incidence and Characterization in Patients on Osteoporosis Treatment: The ICARO Study," *Journal of Bone and Mineral Research*, 21 (2006): 1565–70; and Adami, Isaia, Luisetto, Minisola, Sinigaglia, S. Silvestri, Agnusdei, Gentilella, and Nuti, on behalf of the ICARO Study Group, "Osteoporosis Treatment and Fracture Incidence: The ICARO Longitudinal Study," *Osteoporosis International*, 19 (2008): 1219–23.

70. Stefania Maggi, P. Siviero, S. Gonnelli, C. Caffarelli, G. Gandolini, C. Cisari, M. Rossini, et al., "The Burden of Previous Fractures in Hip Fracture Patients: The Break Study," *Aging: Clinical and Experimental Research*, 23 (2011): 183–86.

Chapter 6. Therapeutic Expansion

1. For data on the use of ERT and progesterones, see Diane K. Wysowski, Linda Golden, and Laurie Burke, "Use of Menopausal Estrogens and Medroxyprogesterone in the United States, 1982–1992," *Obstetrics and Gynecology*, 85 (1995): 6–10.

2. Deborah Grady, S. M. Rubin, D. B. Petitti, C. S. Fox, D. Black, B. Ettinger, V. L. Ernster, and S. R. Cummings, "Hormone Therapy to Prevent Disease and Prolong Life in Postmenopausal Women," *Annals of Internal Medicine*, 117 (1992): 1016–37.

3. American College of Physicians, "Guidelines for Counseling Postmenopausal

Women about Preventive Hormone Therapy," *Annals of Internal Medicine*, 117 (1992): 1038–41.

4. Ibid. For similar recommendations, see Paul E. Belchets, "Hormonal Treatment of Postmenopausal Women," *New England Journal of Medicine*, 330 (1994): 1062–71.

5. Karen K. Steinberg, S. J. Smith, S. B. Thacker, and D. F. Stroup, "Breast Cancer Risk and Duration of Estrogen Use: The Role of Study Design in Meta-Analysis," *Epidemiology*, 5 (1994): 415–21.

6. Eugenia E. Calle, H. L. Miracle-McMahill, M. J. Thun, and C. W. Heath Jr., "Estrogen Replacement Therapy and the Risk of Fatal Colon Cancer in a Prospective Cohort of Postmenopausal Women," *Journal of the National Cancer Institute*, 87 (1995): 517–23. There was insufficient data for 0.3 percent of the group.

7. Jane A. Cauley, D. G. Seeley, K. Ensrud, B. Ettinger, D. Black, and S. R. Cummings, "Estrogen Replacement Therapy and Fractures in Older Women," *Annals of Internal Medicine*, 122 (1995): 9–16.

8. Writing Group for the PEPI Trial, "Effects of Estrogen or Estrogen Regimens on Heart Disease Risk Factors in Postmenopausal Women: The Postmenopausal Estrogen/Progestin Interventions (PEPI) Trial," *JAMA*, 273 (1995): 199–208.

9. Bernadine Healy, "PEPI in Perspective: Good Answers Spawn Pressing Questions [editorial]," *JAMA*, 273 (1995): 240–41.

10. Graham A. Colditz, S. E. Hankinson, D. J. Hunter, W. C. Willett, J. E. Manson, M. J. Stampfer, C. Hennekens, B. Rosner, and F. E. Speizer, "The Use of Estrogens and Progestins and the Risk of Breast Cancer in Postmenopausal Women," *New England Journal of Medicine*, 332 (1995): 1589–93.

11. Nancy E. Davidson, "Hormone-Replacement Therapy—Breast versus Heart versus Bone [editorial]," *New England Journal of Medicine*, 332 (1995): 1638–39.

12. U.S. Congress, Office of Technology Assessment, *Effectiveness and Costs of Osteoporosis Screening and Hormone Replacement Therapy*, 2 vols. (Washington, DC: U.S. Government Printing Office, 1995), vol. 1: 1–4, 50–53 and vol. 2: 5–6, 30, 72–73, 111–18, 176–77, 191–93.

13. D. L. Buick, D. Crook, and R. Horne, "Women's Perceptions of Hormone Replacement Therapy: Risks and Benefits (1980–2002); A Literature Review," *Climacteric*, 8 (2005): 24–35.

14. Lynn M. Meadows, L. A. Mrkonjic, L. E. Lagendyk, and K. M. Petersen, "After the Fall: Women's Views of Fractures in Relation to Bone Health at Midlife," *Women and Health*, 39 (2004): 47–62.

15. Loran M. Salamone, L. R. Pressman, D. G. Seeley, and J. A. Cauley, "Estrogen Replacement Therapy," *Archives of Internal Medicine*, 156 (1996): 1293–97.

16. Katherine M. Newton, A. Z. LaCroix, S. G. Leveille, C. Rutter, N. L. Keenan, and L. A. Anderson, "Women's Beliefs and Decisions about Hormone Replacement Therapy," *Journal of Women's Health*, 6 (1997): 459–65.

17. Ibid.; and K. Backett-Milburn, O. Parry, and N. Mauthner, " 'I'll Worry about That When It Comes Along': Osteoporosis, a Meaningful Issue for Women at Mid-

Life?," *Health Education Research*, 15 (2000): 153–62. For similar expressions by affected females about the role of medical authority in the treatment of osteoporosis, see F. P. M. J. Groeneveld, F. P. Bareman, R. Barentsen, H. J. Dokter, A. C. Drogendijk, and A. W. Hoes, "Determinants of First Prescription of Hormone Replacement Therapy: A Follow-Up Study among 1689 Women Aged 45–60 Years," *Maturitas*, 20 (1994): 81–89; and Mark Garton, David Reid, and Elaine Rennie, "The Climacteric, Osteoporosis, and Hormone Replacement: Views of Women Aged 45–49," *Maturitas*, 21 (1995): 7–15.

18. Herbert Fleisch, "Bisphosphonates: A New Class of Drugs in Diseases of Bone and Calcium Metabolism," in *Bisphosphonates and Tumor Osteolysis*, ed. Kurt W. Brunner, Fleisch, and Hans-Jörg Senn (Berlin: Springer-Verlag, 1989), 16; and T. John Martin, "Herbert André Fleisch, MD, 22 July 1933–15 May 2007," *Osteoporosis International*, 18 (2007): 1019–21. Also see Fleisch, R. G. Russell, B. Simpson, and R. C. Mühlbauer, "Prevention by a Diphosphonate of Immobilization 'Osteoporosis' in Rats," *Nature*, 223 (1969): 211–12; and Fleisch, Russell, and Marion D. Francis, "Diphosphonates Inhibit Hydroxypatite Dissolution In Vitro and Bone Resorption in Tissue Culture and In Vivo," *Science*, 165 (1969): 1262–64.

19. Herbert Fleisch, *Bisphosphonates in Bone Disease: From the Laboratory to the Patient*, 1st ed. (Bern: University of Berne, 1993); and 4th ed. (San Diego: Academic Press, 2000).

20. Elyse Tanouye, "Delicate Balance: Estrogen Study Shifts Ground for Women—and for Drug Firms," *Wall Street Journal*, Jun. 15, 1995.

21. R. P. Heaney and P. D. Saville, "Etidronate Disodium in Postmenopausal Osteoporosis," *Clinical Pharmacology and Therapeutics*, 20 (1976): 593–604.

22. Nelson B. Watts, S. T. Harris, H. K. Genant, R. D. Wasnich, P. D. Miller, R. D. Jackson, A. A. Licata, et al., "Intermittent Cyclical Etidronate Treatment of Postmenopausal Osteoporosis," *New England Journal of Medicine*, 323 (1990): 73–79.

23. Tommy Storm, G. Thamsborg, T. Steiniche, H. K. Genant, and O. H. Sørensen, "Effect of Intermittent Cyclical Etidronate Therapy on Bone Mass and Fracture Rate in Women with Postmenopausal Osteoporosis," *New England Journal of Medicine*, 322 (1990): 1265–71.

24. Louis V. Avioli, "Therapy with Bisphosphonates, Sodium Fluoride, Thiazides, and Vitamin D Metabolites," in *The Osteoporotic Syndrome: Detection, Prevention, and Treatment*, 3rd ed., ed. Avioli (New York: Wiley-Liss, 1993), 155–60.

25. Charles H. Chestnut III, "Drug Therapy: Calcitonin, Bisphosphonates, and Anabolic Steroids," in *Osteoporosis: Etiology, Diagnosis, and Management*, 2nd ed., ed. B. Lawrence Riggs and L. Joseph Melton III (Philadelphia: Lippincott-Raven, 1995), 394–98.

26. C. H. Chestnut III and S. T. Harris, "Short-Term Effect of Alendronate on Bone Mass and Bone Remodeling in Postmenopausal Women," *Osteoporosis International*, 3, Suppl. 3 (1993): S17–S19; Steven T. Harris, B. J. Gertz, H. K. Genant, D. R Eyre, T. T. Survill, J. N. Ventura, J. DeBrock, E. Ricerca, and C. H. Chestnut III, "The Effect of Short Term Treatment with Alendronate on Vertebral Density and Biochemical Markers of Bone Remodeling in Early Postmenopausal Women,"

Journal of Clinical Endocrinology and Metabolism, 76 (1993): 1399–1406; and A. C. Santora, N. H. Bell, C. H. Chestnut III, K. Ensruf, H. K. Genant, R. Grimm, S. T. Harris, et al., "Oral Alendronate Treatment of Bone Loss in Postmenopausal Osteopenic Women," *Journal of Bone and Mineral Research*, 8, Suppl. 1 (1993): S131.

27. Herbert Fleisch, "Introduction," *Osteoporosis International*, 3, Suppl. 3 (1993): S1; and Fleisch, "Bisphosphonates in Osteoporosis: An Introduction," *Osteoporosis International*, 3, Suppl. 3 (1993): S3–S5.

28. Herbert Fleisch, "New Bisphosphonates in Osteoporosis," *Osteoporosis International*, 3, Suppl. 3 (1993): S1, S15–S22.

29. G. A. Rodan, J. G. Seedor, and R. Balena, "Preclinical Pharmacology of Alendronate," *Osteoporosis International*, 3, Suppl. 3 (1993): S7–S12.

30. Chestnut and Harris, "Short-Term Effect of Alendronate," S17–S19.

31. D. M. Black, T. F. Reiss, M. C. Nevitt, J. Cauley, D. Karpf, and S. R. Cummings, "Design of the Fracture Intervention Trial," *Osteoporosis International*, 3, Suppl. 3 (1993): S29–S39.

32. D. C. Anderson, ""Alendronate: Some Remaining Paradoxes," *Osteoporosis International*, 3, Suppl. 3 (1993): S41–S42.

33. Dennis M. Black, S. R. Cummings, D. B. Karpf, J. A. Cauley, D. E. Thompson, M. C. Nevitt, D. C. Bauer, et al., "Randomised Trial of Effect of Alendronate on Risk of Fracture in Women with Existing Vertebral Fractures," *Lancet*, 348 (Dec. 7, 1996): 1535–51. See also Uri A. Liberman, S. R. Weiss, J. Bröll, H. W. Minne, H. Quan, N. H. Bell, J. Rodriguez-Portales, R. W. Downs Jr., J. Dequeker, and M. Favus, "Effect of Oral Alendronate on Bone Mineral Density and the Incidence of Fractures in Postmenopausal Osteoporosis," *New England Journal of Medicine*, 333 (1995): 1437–43.

34. Adrian Phillips, "The Fracture Intervention Trial [letter]," *Lancet*, 349 (Feb. 15, 1997): 505.

35. Dennis M. Black and Steven R. Cummings, "Authors' Reply," *Lancet*, 349 (Feb. 15, 1997): 505–6.

36. Steven R. Cummings, D. M. Black, D. E. Thompson, W. B. Applegate, E. Barrett-Connor, T. A. Musliner, L. Palermo, et al., "Effect of Alendronate on Risk of Fracture in Women with Low Bone Density but without Vertebral Fractures," *JAMA*, 280 (1998): 2077–82.

37. Robert P. Heaney, "Bone Mass, Bone Fragility, and the Decision to Treat," *JAMA*, 280 (1998): 2119–20.

38. See esp. Fiona Godlee, "Absolute Risk Please," *BMJ*, 336 (2008): front matter; and Ray Moynihan, Iona Heath, and David Henry, "Selling Sickness: The Pharmaceutical Industry and Disease Mongering," *BMJ*, 324 (2002): 886–90.

39. Dieter Felsenberg, A. Alenfeld, O. Bock, C. Hammermeister, and W. Gowan, "Placebo-Controlled Multicenter Study of Oral Alendronate in Postmenopausal Osteoporotic Women," *Maturitas*, 31 (1998): 35–43.

40. Peter F. Schneider, M. Fischer, B. Allollo, D. Felsenberg, U. Schröder, J. Semier, and J. R. Ittner, "Alendronate Increases Bone Density and Bone Strength

at the Distal Radius in Postmenopausal Women," *Journal of Bone and Mineral Research*, 14 (1999): 1387–93.

41. H. A. P. Pols, D. Felsenberg, D. A. Hanley, J. Stepán, M. Muñoz-Torres, T. J. Wilkin, G. Qin-sheng, et al., "Multinational, Placebo-Controlled, Randomized Trial of the Effects of Alendronate on Bone Density and Fracture Risk in Postmenopausal Women with Low Bone Mass: Results of the FOSIT Study," *Osteoporosis International*, 9 (1999): 461–68.

42. Peter Selby, "Alendronate Treatment for Osteoporosis: A Review of the Clinical Evidence," *Osteoporosis International*, 6 (1996): 419–26.

43. *Wall Street Journal*, Jun. 15, 1995.

44. Merck & Co., *Annual Report* (1995), 7; *Annual Report* (2002), 19; and *Annual Report* (2003), 18.

45. *Wall Street Journal*, Jun. 15, 1995. The NIH statement in the company's handout was referring to Colditz et al., "Use of Estrogens and Progestins."

46. Nancy Fox Ray, J. K. Chan, M. Thamer, and L. J. Melton III, "Medical Expenditures for the Treatment of Osteoporotic Fractures in the United States in 1995: Report from the National Osteoporosis Foundation," *Journal of Bone and Mineral Research*, 12 (1997): 24–35.

47. L. J. Melton III, M. Thamer, N. F. Ray, J. K. Chan, C. H. Chestnut III, T. A. Einhorn, C. C. Johnston, L. G. Raisz, S. L. Silverman, and E. S. Siris, "Fractures Attributable to Osteoporosis: Report from the National Osteoporosis Foundation," *Journal of Bone and Mineral Research*, 12 (1997): 16–23.

48. National Osteoporosis Foundation, *Physician's Guide to Prevention and Treatment of Osteoporosis* (Washington, DC: National Osteoporosis Foundation, 1998), 1–2.

49. John A. Kanis, David Torgerson, and Cyrus Cooper, "Comparison of the European and USA Practice Guidelines for Osteoporosis," *Trends in Endocrinology and Metabolism*, 11 (2000): 28–32.

50. Pierre D. Delmas, R. Rizzoli, C. Cooper, and J.-Y. Reginster, "Treatment of Patients with Postmenopausal Osteoporosis Is Worthwhile: The Position of the International Osteoporosis Foundation," *Osteoporosis International*, 16 (2005): 1–5.

51. J.-Y. L. Reginster, "Harmonization of Clinical Practice Guidelines for the Prevention and Treatment of Osteoporosis and Osteopenia in Europe: A Difficult Challenge," *Calcified Tissue International*, 59, Suppl. 1 (1996): S24–S29.

52. T. Fujita, "Clinical Guidelines for the Treatment of Osteoporosis in Japan," *Calcified Tissue International*, 59, Suppl. 1 (1996): S34–S37.

53. The NOF annual financial reports do not provide information on contributions by pharmaceutical firms. Three separate inquiries on this subject by this author did not elicit any reply by the NOF.

54. Bernadine P. Healy, "Weak and Feeble Bones No More: The National Osteoporosis Foundation Speaks Out [editorial]," *Journal of Women's Health*, 7 (1998): 1067–68.

55. Donna F. Heinemann, "Osteoporosis: An Overview of the National Osteoporosis Foundation Clinical Practice Guide," *Geriatrics*, 55, no. 5 (2000): 31–36.

56. "Bone Density Measurement—a Systematic Review: A Report from SBU, the Swedish Council on Technology Assessment in Health Care," *Journal of Internal Medicine*, Suppl. 739 (1997): 1–60.

57. British Columbia Office of Health Technology Assessment, *Bone Mineral Density Testing: Does the Evidence Support Its Selective Use in Well Woman?* (Vancouver: University of British Columbia, 1997), Executive Summary.

58. Ibid., Sections 1.1.3 and 1.2.2.

59. Ibid., Sections 1.3.1–1.3.6.

60. Ibid., S2.

61. British Columbia Office of Health Technology Assessment, *Normal Bone Mass, Aging Bodies, Marketing of Fear: Bone Mineral Density Screening of Well Women* (Vancouver: University of British Columbia, 1998), 2–9.

62. Margaret Lock and Patricia Kaufert, "Menopause, Local Biologies, and Cultures of Aging," *American Journal of Human Biology*, 13 (2001): 494–504.

63. Robert Marcus, David Feldman, and Jennifer Kelsey, eds., *Osteoporosis* (San Diego: Academic Press, 1996), xxiii–xxiv; and 2nd ed., 2 vols. (San Diego: Academic Press, 2001).

64. Cyrus Cooper and L. Joseph Melton III, "Magnitude and Impact of Osteoporosis and Fractures," in *Osteoporosis*, 1st ed., ed. Marcus, Feldman, and Kelsey, 419–33.

65. John A. Kanis, *Textbook of Osteoporosis* (Oxford: Blackwell Sciences, 1996), ix, 279–307.

Chapter 7. Osteoporosis Triumphant?

1. Edward G. Jones and Lorne M. Mendell, "Assessing the Decade of the Brain," *Science*, 284 (1999): 739.

2. P. D. Delmas and M. Anderson, "Launch of the Bone and Joint Decade 2000–2010," *Osteoporosis International*, 11 (2000): 95–97.

3. U.S. Senate, 105th Congress, 2nd Session, *Osteoporosis: Prevention, Education, and Research: Hearing Before a Subcommittee of the Committee on Appropriations*, Senate Hearing 105–861 (Washington, DC: U.S. Government Printing Office, 1999).

4. The discussion of the activities of the conference in this and following paragraphs are drawn from "Osteoporosis Prevention, Diagnosis, and Therapy," NIH Consensus Statement, Mar. 27–29, 2000, http://consensus.nih.gov/historical .htm; and NIH Consensus Development Panel on Osteoporosis Prevention, Diagnosis, and Therapy, "Osteoporosis Prevention, Diagnosis, and Therapy," *JAMA*, 285 (2001): 785–95.

5. Elizabeth S. Watkins, *The Estrogen Elixir: A History of Hormone Replacement Therapy in America* (Baltimore: Johns Hopkins University Press, 2007), 262–67.

6. Stephen Hulley, D. Grady, T. Bush, C. Furberg, D. Harrington, B. Riggs, and E. Vittinghoff, "Randomized Trial of Estrogen plus Progestin for Secondary Prevention of Coronary Heart Disease in Postmenopausal Women," *JAMA*, 280 (1998): 605–13.

7. Diana B. Petitte, "Hormone Replacement Therapy and Heart Disease Prevention: Experimentation Trumps Observation [editorial]," *JAMA*, 280 (1998): 650–52.

8. Trudy L. Bush, "Lessons from HERS: The Null and Beyond [editorial]," *Journal of Women's Health*, 7 (1998): 781–83.

9. Watkins, *Estrogen Elixir*, 268–69.

10. Writing Group for the Women's Health Initiative Investigation, "Risks and Benefits of Estrogen plus Progestin in Healthy Postmenopausal Women: Principal Results from the Women's Health Initiative Randomized Controlled Trial," *JAMA*, 286 (2002): 321–33.

11. James V. Lacey Jr., P. J. Mink, J. H. Lubin, M. E. Sherman, T. Troisi, P. Hartge, A. Schatzkin, and C. Schairer, "Menopausal Hormone Replacement Therapy and Risk of Ovarian Cancer," *JAMA*, 286 (2002): 334–41.

12. Watkins, *Estrogen Elixir*, 270.

13. Writing Group, "Risks and Benefits," 321.

14. Eric Levens and R. Stan Williams, "Current Opinions and Understandings of Menopausal Women about Hormone Replacement Therapy (HRT)—the University of Florida Experience," *American Journal of Obstetrics and Gynecology*, 191 (2004): 641–47.

15. National Women's Health Network, *The Truth about Hormone Replacement Therapy: How to Break Free from the Medical Myths of Menopause* (Roseville, CA: Prima, 2002), 145–76, 242–43. See also Christine Laine, "Postmenopausal Hormone Replacement Therapy: How Could We Have Been So Wrong?," *Annals of Internal Medicine*, 137 (2002): 290.

16. For an incisive and illuminating analysis of the WHI and HRT, see Watkins, *Estrogen Elixir*, 264–85.

17. The material in this paragraph is drawn from A. Z. Bluming and Carol Tavris, "Chains of Evidence, Mosaics of Data: Does Estrogen 'Cause' Breast Cancer? How Would We Know?," *Climacteric*, 15 (2012): 531–37. See also Bluming and Tavris, "What Are the Real Risks for Breast Cancer?," *Climacteric*, 15 (2012): 133–38; and Bluming and Tavris, "Hormone Replacement Therapy: Real Concerns and False Alarms," *Cancer Journal*, 15 (2009): 93–104.

18. For an illuminating discussion of the methodological weaknesses of contemporary epidemiology, see Bruce G. Charlton, "Second Thoughts: Attribution of Causation in Epidemiology; Chain or Mosaic?," *Journal of Clinical Epidemiology*, 49 (1996): 105–7.

19. Sally A. Shumaker, C. Legault, S. R. Rapp, L. Thai, R. B. Wallace, J. K. Ockene, S. L. Hendrix, et al., "Estrogen plus Progestin and the Incidence of Dementia and Mild Cognitive Impairment in Postmenopausal Women: The Women's Health Initiative Memory Study; A Randomized Controlled Trial," *JAMA*, 289 (2003): 2651–62.

20. Stephen A. Rapp, M. A. Espeland, S. A. Shumaker, V. W. Henderson, R. L. Brunner, J. E. Manson, M. L. S. Gass, et al., "Effect of Estrogen plus Progestin

on Global Cognitive Function in Postmenopausal Women: The Women's Health Initiative Memory Study; A Randomized Controlled Trial," *JAMA*, 289 (2003): 2663–72.

21. Sylvia Wassertheil-Smoller, S. Hendrix, M. Limacher, G. Heiss, C. Kooperberg, A. Baird, T. Kotchen, et al., "Effect of Estrogen plus Progestin on Stroke in Postmenopausal Women: The Women's Health Initiative; A Randomized Trial," *JAMA*, 289 (2003): 2673–84.

22. Rowan T. Chlebowski, S. L. Hendrix, R. D. Langer, M. L. Stefanick, M. Gass, D. Lane, R. J. Rodabough, et al., "Influence of Estrogen plus Progestin on Breast Cancer and Mammography in Healthy Postmenopausal Women: The Women's Health Initiative Randomized Trial," *JAMA*, 289 (2003): 3243–53.

23. Peter H. Gann and Monica Morrow, "Combined Hormone Therapy and Breast Cancer: A Single-Edged Sword [editorial]," *JAMA*, 289 (2003): 3304–6.

24. Jennifer Hays, J. K. Ockens, R. L. Brunner, J. M. Kotchen, J. Manson, R. E. Patterson, A. K. Aragaki, et al., "Effects of Estrogen plus Progestin on Health-Related Quality of Life," *New England Journal of Medicine*, 348 (2003): 1839–54.

25. Deborah Grady, "Postmenopausal Hormones—Therapy for Symptoms Only," *New England Journal of Medicine*, 348 (2003): 3855–57.

26. Women's Health Initiative Steering Committee, "Effects of Conjugated Equine Estrogen in Postmenopausal Women with Hysterectomy: The Women's Health Initiative Randomized Controlled Trial," *JAMA*, 291 (2004): 1701–12.

27. "Preventive Services Task Force Issues Caution on Combined Hormone Therapy," Electronic Newsletter No. 74, Agency for Healthcare Research and Quality, U.S. Department of Health and Human Services, Oct. 18, 2002, http://archive .ahrq.gov/news/enews/enews74.htm; Heidi D. Nelson, L. L. Humphrey, P. Nygren, S. M. Teutsch, and J. D. Allan, "Postmenopausal Hormone Replacement Therapy: Scientific Review," *JAMA*, 288 (2002): 872–81; Heidi D. Nelson, "Assessing Benefits and Harms of Hormone Replacement Therapy: Clinical Applications," *JAMA*, 288 (2002): 882–84; and U.S. Preventive Services Task Force, "Hormone Therapy for the Prevention of Chronic Conditions in Postmenopausal Women: Recommendations from the U.S. Preventive Services Task Force," *Annals of Internal Medicine*, 142 (2005): 855–60.

28. Watkins, *Estrogen Elixir*, 279–80.

29. U.S. Department of Health and Human Services, *Bone Health and Osteoporosis: A Report of the Surgeon General* (Rockville, MD: U.S. Department of Health and Human Services, Office of the Surgeon General, 2004), 229–30.

30. Ibid., 231–32.

31. U.S. House of Representatives, 102nd Congress, 1st Session, *Women at Midlife: Consumers of Second-Rate Health Care?; Hearing Before the Subcommittee on Housing and Consumer Interests of the Select Committee on Aging . . . May 30, 1991,* Committee Publication 102–814 (Washington, DC: U.S. Government Printing Office, 1991).

32. U.S. Senate, 102nd Congress, 1st Session, *The Role of Menopause and Gender Difference in Aging on the Development of Disease in Mid-Life and Older Women:*

Hearing Before the Subcommittee on Aging of the Committee on Labor and Human Resources . . . April 19, 1991, Senate Hearing 102–1197 (Washington, DC: U.S. Government Printing Office, 1992), 1–2.

33. For an illuminating discussion, see Watkins, *Estrogen Elixir*, 282–85.

34. NIH State-of-the-Sciences Panel, "National Institutes of Health State-of-the Science Conference Statement: Management of Menopause-Related Symptoms," *Archives of Internal Medicine*, 142 (2005): 1003–13.

35. B. E. Christopher Nordin, "Should the Treatment of Osteoporosis Be More Selective?," *Osteoporosis International*, 14 (2003): 99–102.

36. U.S. Department of Health and Human Services, *Bone Health and Osteoporosis*, iii.

37. Ibid., 221.

38. Ibid., 226–37.

39. Agency for Healthcare Research and Quality, *Comparative Effectiveness of Treatments to Prevent Fractures in Men and Women with Low Bone Density or Osteoporosis*, Comparative Effectiveness Review No. 12 (Rockville, MD: Agency for Healthcare Research and Quality, 2007); and Silvina Lewis and George Theodore, "Summary of AHRQ's *Comparative Effectiveness Review of Treatment to Prevent Fractures in Men and Women with Low Bone Density or Osteoporosis*: Update of the 2007 Report," *Journal of Managed Care Pharmacy*, 18, no. 4, Suppl. B (2012): S1–S15. The full statement can be accessed at www.uspreventiveservicestaskforce.org/uspstf10/osteoporosis/osteors.htm.

40. Pierre D. Delmas and Ego Seeman, "Changes in Bone Mineral Density Explain Little of the Reduction in Vertebral or Nonvertebral Fracture Risk with Anti-Resorptive Therapy," *Bone*, 34 (2004): 599–604.

41. S. J. Gallacher and T. Dixon, "Impact of Treatments for Postmenopausal Osteoporosis (Bisphosphonates, Parathyroid Hormone, Strontium Ranelate, and Denosumab) on Bone Quality: A Systematic Review," *Calcified Tissue International*, 87 (2010): 469–84.

42. William R. McClung, P. Geusens, P. D. Miller, H. Zippel, W. G. Bensen, C. Roux, S. Adami, et al., "Effect of Risedronate on the Risk of Hip Fracture in Elderly Women," *New England Journal of Medicine*, 344 (2001): 333–40.

43. Nguyen D. Nguyen, John A. Eisman, and Tuan V. Nguyen, "Anti-Hip Fracture Efficacy of Bisphosphonates: A Bayesian Analysis of Clinical Trials," *Journal of Bone and Mineral Research*, 21 (2006): 340–49.

44. Catherine MacLean, S. Newberry, M. Maglione, M. McMahon, V. Ranganath, M. Suttorp, W. Mojica, et al., "Systematic Review: Comparative Effectiveness of Treatments to Prevent Fractures in Men and Women with Low Bone Density or Osteoporosis," *Annals of Internal Medicine*, 148 (2008): 197–213; and Amir Qaseem, "Pharmacologic Treatment of Low Bone Density or Osteoporosis to Prevent Fractures: A Clinical Practice Guideline from the American College of Physicians," *Annals of Internal Medicine*, 148 (2008): 404–15.

45. Kristine E. Ensrud and John T. Schousboe, "Vertebral Fractures," *New En-

gland Journal of Medicine, 364 (2011): 1634–42; and Norton M. Hadler, *Rethinking Aging: Growing Old and Living Well in an Overtreated Society* (Chapel Hill: University of North Carolina Press, 2011), 121.

46. For a sampling of the literature dealing with clinical trials of antiosteoporotic drugs, see MacLean, "Systematic Review," 209–13; Gallacher and Dixon, "Impact of Treatments," 481–84; and Robert Hopkins, R. Goeree, E. Pullenayegum, J. D. Adachi, A. Papaioannou, F. Xie, and L. Thebane, "The Relative Efficacy of Nine Osteoporosis Medications for Reducing the Rate of Fractures in Post-Menopausal Women," *BMC Musculoskeletal Disorders*, 12 (2011): 209–24.

47. Dennis M. Black, P. D. Delmas, R. Eastell, I. R. Reid, S. Boonen, J. A. Cauley, F. Cosman, et al., "Once-Yearly Zoledronic Acid for Treatment of Postmenopausal Osteoporosis," *New England Journal of Medicine*, 356 (2007): 1809–22.

48. Ibid.

49. Kenneth W. Lyles, C. S. Colón-Emeric, J. S. Magaziner, J. D. Adachi, C. F. Pieper, C. Mautalen, L. Hyldstrup, et al., "Zoledronic Acid and Clinical Fractures and Mortality after Hip Fracture," *New England Journal of Medicine*, 357 (2007): 1799–1809.

50. Karim A. Calis and Frank Pucino, "Zoledronic Acid and Secondary Prevention of Fractures," *New England Journal of Medicine*, 357 (2007): 1861–62.

51. S. W. Blume and J. R. Curtis, "Medical Costs of Osteoporosis in the Elderly Medicare Population," *Osteoporosis International*, 22 (2011): 1835–44.

52. National Osteoporosis Foundation, "Osteoporosis: Review of the Evidence for Prevention, Diagnosis, and Treatment and Cost-Effective Analysis," *Osteoporosis International*, 8, Suppl. 4 (1998): S1–S2, S66–S69.

53. John T. Schousboe, J. A. Nyman, R. L. Kane, and K. E. Ensrud, "Cost-Effectiveness of Alendronate Therapy for Osteopenic Postmenopausal Women," *Annals of Internal Medicine*, 142 (2005): 734–41.

54. The controversy can be followed in Pierre D. Delmas and Ethel S. Siris, "NICE Recommendations for the Prevention of Osteoporotic Fractures in Post-menopausal Women," *Bone*, 42 (2008): 16–18; "The Cost-Effectiveness of Alendronate in the Management of Osteoporosis [editorial]," *Bone*, 42 (2008): 4–13; J. A. Kanis and J. E. Compston, "NICE Continues to Muddy the Waters of Osteoporosis," *Osteoporosis International*, 19 (2008): 1105–7; and E. M. Dennison and C. Cooper, "Where Now With NICE?," *Osteoporosis International*, 22 (2011): 1007–8.

55. A. Cranney, L. Waldegger, I. D. Graham, M. Man-Son-Hing, A. Byszewski, and D. S. Ooi, "Systematic Assessment of the Quality of Osteoporosis Guidelines," *BMC Musculoskeletal Disorders*, 3 (2002): 20–29.

56. Heidi D. Nelson, M. Helfand, S. H. Woolf, and J. D. Allan, "Screening for Postmenopausal Osteoporosis: A Review of the Evidence for the U.S. Preventive Services Task Force," *Annals of Internal Medicine*, 137 (2002): 529–41; and U.S. Preventive Services Task Force, "Screening for Osteoporosis in Postmenopausal Women: Recommendations and Rationale," *Annals of Internal Medicine*, 137 (2002): 526–28.

57. U.S. Preventive Services Task Force, "Screening for Osteoporosis: U.S. Preventive Services Task Force Recommendation Statement," *Annals of Internal Medicine*, 154 (2011): 356–64.

58. Edward S. Leib, E. M. Lewiecki, N. Binkley, and R. C. Hamdy, on behalf of the International Society for Clinical Densitometry, "Official Positions of the International Society for Clinical Densitometry," *Southern Medical Journal*, 97 (2004): 107–10; Lewiecki, D. L. Kendler, G. M. Kiebzak, P. Schmeer, R. L. Prince, G. El-Hajj Fuleihan, and D. Hans, "Special Report on the Official Positions of the International Society for Clinical Densitometry," *Osteoporosis International*, 15 (2004): 779–804; and Lewiecki, C. M. Gordon, S. Baim, M. B. Leonard, N. J. Bishop, M. L. Bianchi, H. J. Kalkwarf, et al., "International Society for Clinical Densitometry 2007 Adult and Pediatric Official Positions," *Bone*, 43 (2008): 1115–21.

59. Nelson B. Watts, "The Fracture Risk Assessment Tool (FRAX): Applications in Clinical Practice," *Journal of Women's Health*, 20 (2011): 525–31; Watts, Bruce Ettinger, and Meryl S. LeBoff, "Perspective: FRAX Facts," *Journal of Bone and Mineral Research*, 24 (2009): 975–79; and Michael R. McClung, "To FRAX or Not to FRAX," *Journal of Bone and Mineral Research*, 27 (2012): 1240–42. For the official positions of the International Society for Clinical Densitometry and the International Osteoporosis Foundation, see their articles in the *Journal of Clinical Densitometry: Assessment of Skeletal Health*, 14 (2011): 171–80, 194–204, 237–39.

60. National Osteoporosis Foundation, *Physician's Guide to Prevention and Treatment of Osteoporosis*, 2nd ed. (Washington, DC: National Osteoporosis Foundation, 2003).

61. National Osteoporosis Foundation, *Physician's Guide to Prevention and Treatment of Osteoporosis*, 3rd ed. (Washington, DC: National Osteoporosis Foundation, 2008), 2–3.

62. Ibid., 6–8.

63. Ibid., 8–9.

64. Ibid., 9, 13–14.

65. Lionel S. Lim, L. J. Hoeksema, and K. Sherin, on behalf of the American College of Preventive Medicine Prevention Practice Committee, "Screening for Osteoporosis in the Adult U.S. Population: ACPM Position Statement on Preventive Practice," *American Journal of Preventive Medicine*, 36 (2009): 366–75.

66. National Osteoporosis Foundation, *Physician's Guide*, 3rd ed., 19. The NOF brought out a revised guide in 2010 that did not substantially alter its 2008 recommendations.

67. Jessica K. Lambert, Mone Zaidi, and Jeffrey I. Mechanick, "Male Osteoporosis: Epidemiology and the Pathogenesis of Aging Bones," *Current Osteoporosis Reports*, 9 (2011): 229–36.

68. Samantha L. Solimeo, "Living With a 'Women's Disease': Risk Appraisal and Management among Men with Osteoporosis," *Journal of Men's Heath*, 8 (2011): 185–91.

69. See, for example, A. Papaioannou, C. C. Kennedy, A. Cranney, G. Hawker, J. P. Brown, S. M. Kaiser, W. D. Leslie, et al., "Risk Factors for Low BMD in Healthy Men Age 50 Years or Older: A Systematic Review," *Osteoporosis International*, 20 (2009): 507–18; Denise L. Orwig, N. Chiles, M. Jones, and M. C. Hochberg, "Osteoporosis in Men: Update 2011," *Rheumatic Disease Clinics of North America*, 37 (2011): 401–14; and J. A. Kanis, G. Bianchi, J. P. Bilezikian, J. M. Kaufman, S. Khosla, E. Orwoll, and E. Seeman, "Towards a Diagnostic and Therapeutic Consensus in Male Osteoporosis," *Osteoporosis International*, 22 (2011): 2789–98.

70. M. Brooke Herndon, L. M. Schwartz, S. Woloshin, and H. G. Welch, "Implications of Expanding Disease Definitions: The Case of Osteoporosis," *Health Affairs*, 26 (2007): 1702–11. The American College of Obstetricians and Gynecologists' clinical guidelines for osteoporosis appeared in the "ACOG Practice Bulletin No. 50," *Obstetrics and Gynecology*, 103 (2004): 203–16.

71. Meghan G. Donaldson, P. M. Cawthon, L. Y. Lui, J. T. Schousboe, K. E. Ensrud, B. C. Taylor, J. A. Cauley, et al., "Estimates of the Proportion of Older White Women Who Would Be Recommended for Pharmacologic Treatment of the New U.S. National Osteoporosis Foundation Guidelines," *Journal of Bone and Mineral Research*, 24 (2009): 675–80.

72. S. D. Berry, D. P. Kiel, M. G. Donaldson, S. R. Cummings, J. A. Kanis, H. Johansson, and E. J. Samelson, "Application of the National Osteoporosis Foundation Guidelines to Postmenopausal Women and Men: The Framingham Osteoporosis Study," *Osteoporosis International*, 21 (2010): 53–60.

73. B. Dawson-Hughes, A. C. Looker, A. N. Tosteson, H. Johansson, J. A. Kanis, and L. J. Melton III, "The Potential Impact of New National Osteoporosis Foundation Guidance on Treatment Patterns," *Osteoporosis International*, 21 (2010): 41–52.

74. B. Dawson-Hughes, "The Potential Impact of the National Osteoporosis Foundation Guidance on Treatment Eligibility in the USA: An Update in NHANES 2005–2008," *Osteoporosis International*, 23 (2012): 811–20.

75. A. C. Looker, B. Dawson-Hughes, A. N. Tosteson, H. Johansson, J. A. Kanis, and L. J. Melton III, "Hip Fracture Risk in Older US Adults by Treatment Eligibility Status Based On New National Osteoporosis Foundation Guidelines," *Osteoporosis International*, 22 (2011): 541–49.

76. Amir Qaseem, V. Snow, P. Shekelle, R. Hopkins Jr., M. A. Forciea, and D. K. Owens, on behalf of the Clinical Efficacy Assessment Subcommittee of the American College of Physicians, "Pharmacologic Treatment of Low Bone Density or Osteoporosis to Prevent Fractures: A Clinical Practice Guideline from the American College of Physicians," *Annals of Internal Medicine*, 149 (2008): 404–15.

77. "Management of Osteoporosis in Postmenopausal Women: 2006 Position Statement of the North American Menopause Society," *Journal of the North American Menopause Society*, 13 (2006): 340–67; "Management of Osteoporosis in Postmenopausal Women: 2010 Position Statement of the North American Menopause Society," *Journal of the North American Menopause Society*, 17 (2010): 25–43; Peter F. Schnatz, "The 2010 North American Menopause Society Position Statement:

Updates on Screening, Prevention, and Management of Postmenopausal Osteoporosis," *Connecticut Medicine*, 73 (2011): 485–87; John C. Stevenson, on behalf of the International Consensus Group on HRT and Regulatory Issues, "HRT, Osteoporosis, and Regulatory Authorities: Quis Custodiet Ipsos Custodes?," *Human Reproduction*, 21 (2006): 1668–71; and Simon Brown, "IMS Updates Its Recommendations on HRT," *Menopause International*, 17 (2011): 75.

78. L. J. Melton III, M. Thamer, N. F. Ray, J. K. Chan, C. H. Chestnut III, T. A. Einhorn, C. C. Johnston, L. G. Raisz, S. L. Silverman, and E. S. Siris, "Fractures Attributable to Osteoporosis: Report from the National Osteoporosis Foundation," *Journal of Bone and Mineral Research*, 12 (1997): 16–23.

79. Katie L. Stone, D. G. Seeley, L. Y. Lui, J. A. Cauley, K. Ensrud, W. S. Browner, M. C. Nevitt, and S. R. Cummings, on behalf of the Osteoporotic Fractures Research Group, "BMD at Multiple Sites and Risk of Fracture of Multiple Types: Long-Term Results from the Study of Osteoporotic Fractures," *Journal of Bone and Mineral Research*, 18 (2003): 1947–54.

80. See S. Kaptoge, L. I. Benevolenskaya, A. K. Bhalla, J. B. Cannata, S. Boonen, J. A. Falch, D. Felsenberg, et al., "Low BMD Is Less Predictive Than Reported Falls for Future Limb Fractures in Women across Europe: Results from the European Prospective Osteoporosis Study," *Bone*, 36 (2005): 387–98; and Fredrick Eklund, A. Nordström, M. Neovius, O. Svensson, and P. Nordström, "Variation in Fracture Rates by Country May Not Be Explained by Differences in Bone Mass," *Calcified Tissue International*, 83 (2009): 10–16.

81. Virginia A. Moyer, on behalf of the U.S. Preventive Services Task Force, "Prevention of Falls in Community-Dwelling Older Adults: U.S. Preventive Services Task Force Recommendation Statement," *Annals of Internal Medicine*, 157 (2012): 197–204.

82. S. Heinrich, K. Rapp, U. Rissmann, C. Becker, and H. H. König, "Cost of Falls in Old Age: A Systematic Review," *Osteoporosis International*, 21 (2010): 891–902; and J. C. Davis, M. C. Robertson, M. C. Ashe, T. Liu-Ambrose, K. M. Khan, and C. A. Marra, "International Comparison of Cost of Falls in Older Adults Living in the Community: A Systematic Review," *Osteoporosis International*, 21 (2010): 1295–1306.

83. For examples, see Rosanne M. Leipzig, Robert G. Cumming, and Mary E. Tinette, "Drugs and Falls in Older People: A Systematic Review and Meta-Analysis; I. Psychotropic Drugs; II. Cardiac and Analgesic Drugs," *Journal of the American Geriatrics Society*, 37 (1999): 30–50; Barbara A. Liu, G. Anderson, N. Mittmann, T. To, T. Axcell, and N. Shear, "Use of Selective Serotonin-Reuptake Inhibitors or Tricyclic Antidepressants and the Risk of Hip Fractures in Elderly People," *Lancet*, 351 (May 2, 1998): 1303–7; M. W. M. van den Brand, S. Pouwels, M. M. Samson, T. P. van Staa, B. Thio, C. Cooper, H. G. Leufkens, A. C. Egberts, H. J. Verhaar, and F. de Vries, "Use of Anti-Depressants and the Risk of Fracture of the Hip or Femur," *Osteoporosis International*, 20 (2009): 1705–13; Chunliu Zhan, Judith Sangl, Arlene S. Bierman, Marlene R. Miller, Bruce Friedman, Steve W. Wickizer, and Gregg S. Meyer, "Potentially Inappropriate Medication Use in the Community-

Dwelling Elderly: Findings from the 1996 Medical Expenditure Panel Survey," *JAMA*, 286 (2001): 2823–29; and Jerry Ahorn, "Improving Drug Use in Elderly Patients: Getting to the Next Level," *JAMA*, 286 (2001): 2866–68.

84. Victoria L. Tseng, "Risk of Fractures following Cataract Surgery in Medicare Beneficiaries," *JAMA*, 308 (2012): 493–501.

85. For a review of RCTs dealing with falls and their prevention, see M. K. Karlsson, H. Magnusson, T. von Schewelov, and B. E. Rosengren, "Prevention of Falls in the Elderly—a Review," *Osteoporosis International*, 24 (2013): 747–62.

86. American Geriatrics Society, British Geriatrics Society, and American Academy of Orthopedic Surgeons Panel on Falls Prevention, "Guidelines for the Prevention of Falls in Older Persons," *Journal of the American Geriatrics Society*, 49 (2001): 664–72.

87. "Summary of the Updated American Geriatrics Society / British Geriatrics Society Clinical Practice Guideline for Prevention of Falls in Older Persons," *Journal of the American Geriatrics Society*, 59 (2011): 148–57.

88. Judy A. Stevens, *A CDC Compendium of Effective Fall Interventions: What Works for Community-Dwelling Older Adults*, 2nd ed. (Atlanta: Centers for Disease Control and Prevention, National Center for Injury Prevention and Control, 2010); National Institute on Aging, "AgePage: Falls and Fractures," www.nia.nih.gov /health/publication/falls-and-fractures [accessed Jun. 27, 2013]; and Moyer, "Prevention of Falls."

89. A. E. Salter, K. M. Khan, M. G. Donaldson, J. C. Davis, J. Buchanan, R. B. Abu-Laban, W. L. Cook, S. R. Lord, and H. A. McKay, "Community-Dwelling Seniors Who Present to the Emergency Department with a Fall Do Not Receive Guideline Care and Their Fall Risk Profile Worsens Significantly: A 6-Month Prospective Study," *Osteoporosis International*, 17 (2006): 672–83.

90. Teppo Järvinen, H. Sievänen, K. M. Khan, A. Heinonen, and P. Kannus, "Shifting the Focus in Fracture Prevention from Osteoporosis to Falls," *BMJ*, 336 (2008): 124–26.

91. Salvatore L. Ruggiero, B. Mehrotra, T. J. Rosenberg, and S. L. Engroff, "Osteonecrosis of the Jaw Associated with the Use of Bisphosphonates: A Review of 63 Cases," *Journal of Oral and Maxillofacial Surgery*, 62 (2004): 527–34. See also Robert E. Marx, "Pamidronate (Aredia) and Zolendronate (Zometa) Induced Avascular Necrosis of the Jaws: A Growing Epidemic," *Journal of Oral and Maxillofacial Surgery*, 61 (2003): 1115–18; and Matthew C. Farrugia, D. J. Summerlin, E. Krowiak, T. Huntley, S. Freeman, R. Borrowdale, and C. Tomich, "Osteonecrosis of the Mandible or Maxilla Associated with the Use of New Generation Bisphosphonates," *Laryngoscope*, 116 (2006): 115–20.

92. Brian G. Durie, Michael Katz, and John Crowley, "Osteonecrosis of the Jaw and Bisphosphonates [letter]," *New England Journal of Medicine*, 353 (2005): 99–100.

93. Clarita V. Odvina, J. E. Zerwekh, D. S. Rao, N. Maalouf, F. A. Gottschalk, and C. Y. Pak, "Severely Suppressed Bone Turnover: A Potential Complication of

Alendronate Therapy," *Journal of Clinical Endocrinology and Metabolism*, 90 (2005): 1294–1301.

94. Susan M. Ott, "Long-Term Safety of Bisphosphonates [editorial]," *Journal of Clinical Endocrinology and Metabolism*, 90 (2005): 1897–99.

95. Sundeep Khosla, D. Burr, J. Cauley, D. W. Dempster, P. R. Ebeling, D. Felsenberg, R. F. Gagel, et al., "Bisphosphonate-Associated Osteonecrosis of the Jaw: Report of a Task Force of the American Society for Bone and Mineral Research," *Journal of Bone and Mineral Research*, 22 (2007): 1479–91.

96. S. K. Goh, K. Y. Yang, J. S. Koh, M. K. Wong, S. Y. Chua, D. T. Chua, and T. S. Howe, "Subtrochanteric Insufficiency Fractures in Patients on Alendronate Therapy," *Journal of Bone and Joint Surgery* (British ed.), 89-B (2007): 349–53.

97. Arkan S. Sayed-Noor and Göran O. Sjödén, "Subtrochanteric Displaced Insufficiency Fracture after Long-Term Alendronate Therapy—a Case Report," *Acta Orthopaedica*, 79 (2008): 565–67; Andrew S. Neviaser, J. M. Lane, B. A. Lenart, F. Edobor-Osula, and D. G. Lorich, "Low-Energy Femoral Shaft Fractures Associated with Alendronate Use," *Journal of Orthopaedic Trauma*, 22 (2008): 346–50; and Joon Kiong Lee, "Bilateral Atypical Femoral Diaphyseal Fractures in a Patient Treated with Alendronate Sodium," *International Journal of Rheumatic Diseases*, 12 (2009): 149–54.

98. Jennifer P. Schneider, "Bisphosphonates and Low-Impact Femoral Fractures: Current Evidence on Alendronate Fracture Risk," *Geriatrics*, 64 (2009): 18–23. See also Atul F. Kamath, "Current Controversies in Bisphosphonate Therapy," *Orthopedics*, 12 (2009): 473–75.

99. Dennis M. Black, M. P. Kelly, H. K. Genant, L. Palermo, R. Eastell, C. Bucci-Rechtweg, J. Cauley, et al., "Bisphosphonates and Fractures of the Subtrochanteric or Diaphyseal Femur," *New England Journal of Medicine*, 362 (2010): 1761–71.

100. Laura Y. Park-Wylie, M. M. Mamdani, D. N. Juurlink, G. A. Hawker, N. Gunraj, P. C. Austin, D. B. Whelan, P. J. Weiler, and A. Laupacis, "Bisphosphonate Use and the Risk of Subtrochanteric or Femoral Shaft Fractures in Older Women," *JAMA*, 305 (2011): 783–89. See also Jörg Schilcher, Karl Michaëlsson, and Per Aspenberg, "Bisphosphonate Use and Atypical Fractures of the Femoral Shaft," *New England Journal of Medicine*, 364 (2011): 1728–37.

101. Bo Abrahamsen, Pia Eiken, and Richard Eastell, "Cumulative Alendronate Dose and the Long-Term Absolute Risk of Subtrochanteric and Diaphyseal Femur Fractures: A Register-Based National Cohort Analysis," *Journal of Clinical Endocrinology and Metabolism*, 95 (2010), 5258–65.

102. M. Pazianes, B. Abrahamsen, Y. Wang, and R. G. Russell, "Incidence of Fractures of the Femur, including Subtrochanteric, up to 8 Years since Initiation of Oral Bisphosphonate Therapy: A Register-Based Cohort Study Using the US MarketScan Claims Databases," *Osteoporosis International*, 23 (2012): 2873–84.

103. The three trial were the Fosamax Long-Term Extension (FLEX), the Reclast Health Outcomes and Reduced Incidence with Zoledronic Acid Once Yearly–

Pivotal Fracture Trial (HORIZON-PFT), and the Actonel Vertebral Efficacy with Risedronate Therapy–Multinational Trial (VERT-MN) Extension.

104. Marcea Whitaker, J. Guo, T. Kehoe, and G. Benson, "Bisphosphonates for Osteoporosis—Where Do We Go from Here?," *New England Journal of Medicine*, 366 (2012): 2048–51.

105. Dennis M. Black, D. C. Bauer, A. V. Schwartz, S. R. Cummings, and C. J. Rosen, "Continuing Bisphosphonate Treatment for Osteoporosis—for Whom and for How Long?," *New England Journal of Medicine*, 2051–53.

106. Ego Seeman, "To Stop or Not to Stop, That Is the Question," *Osteoporosis International*, 20 (2009): 187–95.

107. Ian R. Reid, "Bisphosphonates in the Treatment of Osteoporosis: A Review of Their Contribution and Controversies." *Skeletal Radiology*, 40 (2011): 1191–96.

108. For examples of the debate over adverse events, see Bo Abrahamsen, "Adverse Effects of Bisphosphonates," *Calcified Tissue International*, 86 (2010): 421–35; Andrew J. Laster and S. Bobo Tanner, "Duration of Treatment in Postmenopausal Osteoporosis: How Long to Treat and What Are the Consequences of Cessation of Treatment?," *Rheumatic Disease Clinics of North America*, 37 (2011): 323–36; and Je Zhang, Kenneth G. Saag, and Jeffrey R. Curtis, "Long Term Safety Concerns of Antiresorptive Therapy," *Rheumatic Disease Clinics of North America*, 37 (2011): 387–400.

109. John C. Stevenson, "Prevention of Osteoporosis: One Step Forward, Two Steps Back," *Menopause International*, 17 (2011): 137–41.

110. Russel Burge, B. Dawson-Hughes, D. H. Solomon, J. B. Wong, A. King, and A. Tosteson, "Incidence and Economic Burden of Osteoporosis-Related Fractures in the United States, 2005–2025," *Journal of Bone and Mineral Research*, 22 (2007): 465–75.

111. Pekka Kannus, S. Niemi, J. Parkkari, M. Palvanen, I. Vuori, and M. Järvinen, "Nationwide Decline in Incidence of Hip Fracture," *Journal of Bone and Mineral Research*, 21 (2006): 1836–38.

112. L. Joseph Melton III, E. J. Atkinson, and R. Madhok, "Downturn in Hip Fracture Incidence," *Public Health Reports*, 111 (1996): 146–50; and Melton, A. E. Kearns, Atkinson, M. E. Bolander, S. J. Achenbach, T. M. Huddleston, T. M. Therneau, and C. L. Leibson, "Secular Trends in Hip Fracture Incidence and Recurrence," *Osteoporosis International*, 20 (2009): 687–94.

113. T. Chevalley, E. Guilley, F. R. Herrmann, P. Hoffmeyer, C. H. Rapin, and R. Rizzoli, "Incidence of Hip Fracture over a 10-Year Period (1991–2000): Reversal of a Secular Trend," *Bone*, 40 (2007): 1284–89.

114. Klaas A. Hartholt, C. Oudshoorn, S. M. Zielinski, P. T. Burgers, M. J. Panneman, E. F. van Beeck, P. Patka, and T. J. van der Cammen, "The Epidemic of Hip Fractures: Are We on the Right Track?," *PLoS ONE*, 6 (2011): 1–6 (e2227).

115. William D. Leslie, S. O'Donnell, S. Jean, C. Lagacé, P. Walsh, C. Bancej, S. Morin, D. A. Hanley, A. Papaioannou, and the Osteoporosis Surveillance Working Group, "Trends in Hip Fracture Rates in Canada," *JAMA*, 302 (2009): 883–89. For multinational studies of trends in osteoporotic fracture rates, see C. Cooper,

Z. A. Cole, C. R. Holroyd, S. C. Earl, N. C. Harvey, E. M. Dennison, L. J. Melton III, S. R. Cummings, J. A. Kanis, and the IOF CSA [Committee of Scientific Advisors] Working Group on Fracture Epidemiology, "Secular Trends in the Incidence of Hip and Other Osteoporotic Fractures," *Osteoporosis International*, 22 (2011): 1277–88; and J. A. Kanis, A. Odén, E. V. McCloskey, H. Johansson, D. A. Wahl, C. Cooper, and the IOF Working Group on Epidemiology and Quality of Life, "A Systematic Review of Hip Fracture Incidence and Probability of Fracture Worldwide," *Osteoporosis International*, 23 (2012): 2239–56.

116. E. Cassell and A. Clapperton, "A Decreasing Trend in All-Related Hip Fracture Incidence in Victoria, Australia," *Osteoporosis International*, 24 (2012): 99–109.

117. Barbara Lukert, S. Satram-Hoang, S. Wade, M. Anthony, G. Gao, and R. Downs, "Physician Differences in Managing Postmenopausal Osteoporosis: Results from the POSSIBLE US™ Treatment Registry Study," *Drugs & Aging*, 28 (2011): 713–27.

118. Peter F. Schnatz, K. A. Marakovits, M. Dubois, and D. M. O'Sullivan, "Osteoporosis Screening and Treatment Guidelines: Are They Being Followed?," *Journal of the North American Menopause Society*, 18 (2011): 1072–78.

119. Heike A. Bischhoff-Ferrari, W. C. Willett, J. B. Wong, E. Giovannucci, T. Dietrich, and B. Dawson-Hughes, "Fracture Prevention with Vitamin D Supplementation: A Meta-Analysis of Randomized Controlled Trials," *JAMA*, 293 (2005): 2257–64.

120. U.S. Preventive Services Task Force, *Vitamin D and Calcium Supplementation to Prevent Cancer and Osteoporotic Fractures in Adults: Draft Recommendation Statement*, AHRQ Publication No. 12-05163-EF-2, www.uspreventiveservicestaskforce .org/uspstf12/vitamind/draftrecvitd.htm.

121. Stuart H. Ralston, Roberto Civitelli, and Heike Bischoff-Ferrari, "Casting New Light on the Sunshine Vitamin," *Calcified Tissue International*, 92 (2013): 75–76.

122. Kerrie M. Sanders, Geoffrey C. Nicholson, and Peter R. Ebeling, "Is High Dose Vitamin D Harmful?," *Calcified Tissue International*, 92 (2013): 191–206.

123. John S. Sampalis, J. D. Adachi, E. Rampakakis, J. Vaillancourt, A. Karellis, and C. Kindundu, "Long-Term Impact of Adherence to Oral Bisphosphonates on Osteoporotic Fracture Incidence," *Journal of Bone and Mineral Research*, 27 (2012): 202–10.

124. The following are representative samples: C. G. Unson, E. Siccion, J. Gaztambide, S. Gaztambide, P. Mahoney Trella, and K. Prestwood, "Nonadherence and Osteoporosis Treatment Preferences of Older Women: A Qualitative Study," *Journal of Women's Health*, 12 (2003): 1037–45; Lynn M. Meadows, L. A. Mrkonjic, L. E. Lagendyk, and K. M. Petersen, "After the Fall: Women's Views of Fractures in Relation to Bone Health at Midlife," *Women and Health*, 39 (2004): 47–62; Kathleen M. Mazor, S. Velten, S. E. Andrade, and R. A. Yood, "Older Women's Views about Prescription Osteoporosis Medication: A Cross-Sectional, Qualitative Study," *Drugs & Aging*, 27 (2010): 999–1008; Karen Roush, "Prevention and Treatment of

Osteoporosis in Postmenopausal Women: A Review," *American Journal of Nursing*, 111 (2011): 26–37; J. E. M. Sale, M. A. Gignac, L. Frankel, G. Hawker, D. Beaton, V. Elliot-Gibson, and E. Bogoch, "Patients Reject the Concept of Fragility Fracture— a New Understanding Based on Fracture Patients' Communication," *Osteoporosis International*, 23 (2012): 2829–34; and Sampalis, "Long-Term Impact," 202–10.

125. Meadows et al., "After the Fall," 48.

126. Susan M. Love, *Dr. Susan Love's Hormone Book* (New York: Three Rivers Press, 1997), 74–96. See also Barbara Seaman, *The Greatest Experiment Ever Performed on Women: Exploding the Estrogen Myth* (New York: Hyperion, 2003).

127. National Women's Health Network, *Truth about Hormone Replacement Therapy*, 146–55.

128. Gillian Sanson, *The Myth of Osteoporosis* (Ann Arbor, MI: MCD Century Publications, 2003).

129. Norton M. Hadler, *Rethinking Aging: Growing Old and Living Well in an Overtreated Society* (Chapel Hill: University of North Carolina Press, 2011), 111.

130. Charlotte I. Salter, A. Howe, L. McDaid, J. Blacklock, E. Lenaghan, and L. Shepstone, "Risk, Significance, and Biomedicalisation of a New Population: Older Women's Experience of Osteoporosis Screening," *Social Science & Medicine*, 73 (2011): 808–15.

131. Fiona Godlee, "Absolute Risk Please," *BMJ*, 336 (2008): front matter.

132. Pablo Alonso-Coello, A. L. García-Franco, G. Guyatt, and R. Moynihan, "Drugs for Pre-Osteoporosis: Prevention or Disease Mongering?," *BMJ*, 336 (2008): 126–29.

133. Marcia Angell, "Industry-Sponsored Clinical Research: A Broken System," *JAMA*, 300 (2008): 1069–71. For elaborations on this development, see Marcia Angell, *The Truth about the Drug Companies: How They Deceive Us and What to Do about It* (New York: Random House, 2004); Ray Moynihan, Iona Health, and David Henry, "Selling Sickness: The Pharmaceutical Industry and Disease Mongering," *BMJ*, 324 (2002): 886–90; Bodil Als-Nielson, W. Chen, C. Gluud, and L. L. Kjaergard, "Association of Funding and Conclusions in Randomized Drug Trials: A Reflection of Treatment Effects or Adverse Events?," *JAMA*, 290 (2003): 921–28; Joseph S. Ross, "Guest Authorship and Ghostwriting in Publications Related to Rofecoxib: A Case Study of Industry Documents from Rofecoxib Litigation," *JAMA*, 299 (2008): 1800–1812; Catherine D. DeAngelis and Phil B. Fontanarosa, "Impugning the Integrity of Medical Science: The Adverse Effects of Industry Influence," *JAMA*, 299 (2008): 1833–35; O. Bruyere, J. A. Kanis, M. E. Ibar-Abadie, N. Alsayed, M. L. Brandi, N. Burlet, D. L. Cahall, et al., "The Need for a Transparent, Ethical, and Successful Relationship between Academic Scientists and the Pharmaceutical Industry: A View of the Group for the Respect of Ethics and Excellence in Science (GREES)," *Osteoporosis International*, 21 (2010): 713–22; Jerry Avorn, *Powerful Medicines: The Benefits, Risks, and Costs of Prescription Drugs* (New York: Alfred A. Knopf, 2004); Ray Mohnihan, *Selling Sickness: How the World's Biggest Pharmaceutical Companies Are Turning Us All into Patients* (New York: Nation

Books, 2005); Jerome P. Kassirer, *On the Take: How America's Complicity with Big Business Can Endanger Your Health* (New York: Oxford University Press, 2005); and Thomas J. Nesi, *Poison Pills: The Untold Story of the Vioxx Dug Scandal* (New York: Thomas Dunne Books, 2008).

134. Florence T. Bourgeois, Srinivas Murthy, and Kenneth D. Mandl, "Outcome Reporting among Drug Trials Registered in ClinicalTrials.gov," *Annals of Internal Medicine*, 153 (2010): 158–66.

135. Ben Goldacre's *Bad Pharma: How Drug Companies Mislead Doctors and Harm Patients* (New York: Faber & Faber, 2012) provides an extensive analysis of this subject.

136. Fuller Albright and Edward C. Reifenstein Jr., *The Parathyroid Glands and Metabolic Bone Disease: Selected Studies* (Baltimore: Williams & Wilkins, 1948), 1.

137. Paul B. Beeson, "Changes in Medical Therapy during the Past Half Century," *Medicine*, 59 (1980): 79–85.

138. Kerr L. White, "Restructuring the International Classification of Diseases: Need for a New Paradigm," *Journal of Family Practice*, 21 (1985): 17–19. See also Geoffrey C. Bowker and Susan L. Star, *Sorting Things Out: Classification and Its Consequences* (Cambridge, MA: MIT Press, 1999), 107–33.

139. David L. Sackett, "The Arrogance of Preventive Medicine," *Canadian Medical Association Journal*, 167 (2002): 363. See esp. R. Brian Haynes, David L. Sackett, Gordo H. Guyatt, and Peter Tugwell, *Clinical Epidemiology: How to Do Clinical Practice Research*, 3rd ed. (Philadelphia: Lippincott Williams & Wilkins, 2005). The first two editions (1985 and 1991) listed Sackett as the first author.

140. World Health Organization, *Assessment of Fracture Risk and Its Application to Screening for Postmenopausal Osteoporosis: Report of a WHO Study Group*, WHO Technical Report No. 843 (Geneva: World Health Organization, 1994), 5.

INDEX